International Federation of Automatic Control

SAFETY OF COMPUTER CONTROL SYSTEMS, 1991

Safety, Security and Reliability of Computer Based Systems

IFAC Symposia Series, 1991. Number 8

IFAC SYMPOSIA SERIES

Editor-in-Chief

Janos Gertler, Department of Electrical Engineering,
George Mason University, Fairfax, Virginia 22030, USA

Other IFAC Publications

AUTOMATICA

the journal of IFAC, the International Federation of Automatic Control
Editor-in-Chief: G.S. Axelby, 211 Coronet Drive, North Linthicum,
Maryland 21090, USA

IFAC WORKSHOP SERIES

Editor-in-Chief: Pieter Eykhoff, University of Technology, NL-5600 MB Eindhoven,
The Netherlands

Full list of IFAC Publications appears at the end of this volume

SAFETY OF COMPUTER CONTROL SYSTEMS 1991 (SAFECOMP'91)

Safety, Security and Reliability of Computer Based Systems

Proceedings of the IFAC/IFIP/EWICS/SRE Symposium,
Trondheim, Norway, 30 October - 1 November 1991

Edited by

J. F. LINDEBERG

DELAB, Trondheim, Norway

Published for the

INTERNATIONAL FEDERATION OF AUTOMATIC CONTROL

by

PERGAMON PRESS

OXFORD · NEW YORK · SEOUL · TOKYO

UK	Pergamon Press plc, Headington Hill Hall, Oxford OX3 0BW, England
USA	Pergamon Press, Inc., 395 Saw Mill River Road, Elmsford, New York 10523, USA
KOREA	Pergamon Press Korea, KPO Box 315, Seoul 110-603, Korea
JAPAN	Pergamon Press, 8th Floor, Matsuoka Central Building, 1-7-1 Nishi-Shinjuku, Shinjuku-ku, Tokyo 160, Japan

First edition 1991

Library of Congress Cataloguing in Publication Data

Safety of computer control systems, 1991 (SAFECOMP'91) : safety, security, and reliability : proceedings of the IFAC/IFIP/EWICS/SRE Symposium, Trondheim, Norway, 30 October - 1 November 1991/ edited by J.F. Lindeberg.
p. cm. -- (IFAC symposia series : 1991, no. 8)
Proceedings of the IFAC/IFIP/EWICS/SRE Symposium on Safety of Computer Control Systems.
1. Automatic control -- Reliability -- Congresses. 2. Computers -- Reliability -- Congresses. 3. Computer security -- Congresses.
I. Lindeberg, J.F. II. IFAC/IFIP/EWICS/SRE Symposium on Safety of Computer Control Systems (1991 : Trondheim, Norway)
III. International Federation of Automatic Control. IV Series.
TJ212.2.S239 1991   670.42'7 -- dc20   91-27627

British Library Cataloguing in Publication Data

Safety of Computer Control Systems. Conference (1991 : Trondheim, Norway)
Safety of computer control systems 1991 (Safecomp'91) -
(IFAC symposia series; 1991, no.8)
I. Title II. Lindeberg, J.F. III. Series
670.427

ISBN 9780080416977

These proceedings were reproduced by means of the photo-offset process using the manuscripts supplied by the authors of the different papers. The manuscripts have been typed using different typewriters and typefaces. The lay-out, figures and tables of some papers did not agree completely with the standard requirements: consequently the reproduction does not display complete uniformity. To ensure rapid publication this discrepancy could not be changed: nor could the English be checked completely. Therefore, the readers are asked to excuse any deficiencies of this publication which may be due to the above mentioned reasons.

The Editor

Transferred to digital print 2009
Printed and bound in Great Britain by CPI Antony Rowe, Chippenham and Eastbourne

IFAC/IFIP/EWICS/SRE SYMPOSIUM ON SAFETY OF COMPUTER CONTROL SYSTEMS (SAFECOMP'91)

Safety, Security and Reliability of Computer Based Systems

Sponsored by
International Federation of Automatic Control (IFAC)

Co-sponsored by
International Federation for Information Processing (IFIP)
- TC 10 Computer Systems Technology WG 5-4 Computerized Process Control
Society of Reliability Engineers (SRE) - Scandinavian Chapter
The University of Trondheim, The Norwegian Institute of Technology (NTH)
The Norwegian Society of Chartered Engineers (NIF)
The Foundation for Scientific, Industrial and Technical Research at the Norwegian Institute of
    Technology (SINTEF)

Intiated by
European Workshop on Industrial Computer Systems (EWICS)
- TC7 on Reliability, Safety and Security

International Programme Committee

J.F. Lindeberg (N)   (Chairman)
R. Bloomfield (UK)
S. Bologna (I)
L. Boullart (B)
G. Dahll (N)
B.K. Daniels (UK)
H. de Kroes (NL)
W. Ehrenberger (D)
H. Frey (CH)
C. Galivel (F)
R. Genser (A)
J. Gorski (PL)
S.L. Hansen (DK)
S.J. Knapskog (N)
F. Koornneef (NL)

R. Lauber (D)
N. Leveson (USA)
S. Lydersen (N)
L. Motus (SU)
E. Pilaud (F)
J.M.A. Rata (F)
M. Rausand (N)
F. Redmill (UK)
G.L. Reijns (NL)
M. Rodd (UK)
B. Runge (DK)
L. Simoncini (I)
B. Sterner (S)
A. Toola (SF)
R. Yunker (USA)

National Organizing Committee

S. Lydersen   (Chairman)
M. Balle
J.F. Lindeberg

Previous 'SAFECOMP' Proceedings published by Pergamon Press

DANIELS:
Safety of Computer Control Systems, 1990
(Safety, Security and Reliability Related Computers for the 1990's)
October/November 1990, Gatwick, UK

GENSER et al:
Safety of Computer Control Systems, 1989
December 1989, Vienna, AUSTRIA

EHRENBERGER:
Safety of Computer Control Systems, 1988
(Safety Related Computers in an Expanding Market)
November 1988, Fulda, GERMANY

QUIRK:
Safety of Computer Control Systems, 1986
(Trends in Safe Real Time Computer Systems)
October 1986, Sarlat, FRANCE

QUIRK:
Safety of Computer Control Systems, 1985
(Achieving Safe Real Time Computer Systems)
October 1985, Como, ITALY

BAYLIS:
Safety of Computer Control Systems, 1983
September 1983, Cambridge, UK

LAUBER:
Safety of Computer Control Systems, 1979
May 1979, Stuttgart, GERMANY

<u>NOTE</u>

Due to the material not being made available,
proceedings of the SAFECOMP Workshop held
in Indiana, USA in 1982 were not published.

PREFACE

There exists today a demand for ever more and increasingly complex safety systems and safety-related systems. In addition there is an increased expectation that safety precautions shall not impede reliability and availability of technical operations. Furthermore, the concept of safety now includes risks to life and limb, damage to environment (flora, fauna and landscape) and property.

Under these boundary conditions it is required that acceptable risk levels shall be maintained. The perception by the public of what are acceptable risk levels is changing into the restrictive direction. This places extreme requirements on management of systems development and operations, and on technical solutions.

In this context SAFECOMP'91 contributes to exposing important technological and technical aspects necessary for maintaining an acceptable safety level in the face of threats from numerous sources and for many reasons - from plain mistakes, forgetfulness, negligence, to malevolent actions.

In twenty nine contributed papers, specification, development, analyses, verification and validation of hardware and software are covered. In addition, reliability, availability, security and timing problems are covered.

As Chairman of the International Program Committee (IPC) for SAFECOMP'91 and Editor of this proceedings I would like to thank all members of the Committee for their advice and assistance in constructing the Program. The Program Committee included many members of the European Workshop on Industrial Computer Systems Technical Committee 7 (Safety Security and Reliability), who have provided the core of each SAFECOMP IPC since the first meeting in 1979. The Committee was grateful for the assistance of a number of additional members who are key workers in this field.

My thanks must also go to session Chairmen, the authors and co-authors of the papers, and the members of the National Organizing Committee.

Johan F. Lindeberg
SINTEF DELAB
N-7034 Trondheim
NORWAY

CONTENTS

ASSESSMENT II

SPECIFICATION AND DEVELOPMENT

SECURITY

VERIFICATION AND VALIDATION

MODELS

TOWARDS A COMMON SAFETY DESCRIPTION MODEL

R. E. Bloomfield*, J. H. Cheng* and J. Gorski**

*Adelard, London, UK
**Institute of Informatics, Technical University of Gdansk, Poland

Abstract In order to apply safety analysis techniques in an integrated fashion, the Common Safety Description Model (CSDM) is developed to provide a formal semantics to such techniques. The motivation and objectives of CSDM are discussed; a theory of events that incorporates causality and timing is presented with its applications to safety analysis techniques.

Keywords Safety analysis, semantics, event, causality, timing, fault tree.

1 Introduction

Safety analysis is an important task fulfilled by both assessors and developers and involves application of a number of safety analysis and assessment methods. It is desirable that this is done in such a way that the final effect well exceeds the simple sum of effects of the applied methods. To achieve this, a framework for integration of the results from different methods is necessary. A prerequisite for such integration is an understanding of the methods; we need a model for capturing the precise semantics of the methods, and such a model is named the *Common Safety Description Model* (CSDM).

The structure of this paper is as follows. Section 2 gives the rationale of our approach; Section 3 presents a theory of events, which we call the *event model*; Section 4 demonstrates applications of the theory to a few hazard analysis techniques; Section 5 concludes the paper.

2 Rationale of CSDM

2.1 Limitations of Existing Techniques

The existing, widely used safety analysis techniques appear to have limitations in the following aspects:

semantics usually, such a technique or method is informally defined and therefore open to different interpretations; for example, the OR-gates in fault tree analysis have different meanings, as shown in [Vesely 81] and [HSE 87];

timing existing techniques do not seem able to deal with timing constraint or requirement in a satisfactory way, while such information is often crucial in analysing behaviour of systems.

CSDM aims to remedy such limitations by a theory that is able to provide formal semantics to safety analysis techniques and to cover essential notions of safety analysis.

2.2 Desirable Features

In analysing behaviour of systems, the following issues seem particularly relevant:

event seems central to all kinds of safety analysis; nonetheless, the term in different hazard analysis methods may have different meanings;

causality is a basic relation among events sought for during safety analysis and tells us how some events contribute to other events;

timing information is often crucial in analysing behaviour of systems and hence must be accommodated;

nondeterminism identification of which shows possible ways of leading to accidents;

generalization groups events into classes and treats a whole class as a single event;

compositionality allows one to deal with a composite construct in terms of its immediate components.

Our theory of event, the event model, should have a proper treatment of the issues above.

3 Proposed Event Model

The event model is based on the ideas from [Murphy 90], where a theory following Winskel's *event structures* is presented to handle real-timed concurrent systems. A distinguishing difference is that we adopt a linear time model whereas in [Murphy 90] a branching time model is used.

3.1 Events, Tasks and Transitions

We first make a distinction between an event and an instance of an event. The former is essentially a name to denote an act, such as "cycling to work" and "having lunch", whereas the latter denotes a particular occurrence of an event, such as "having lunch on the 21st June 1991". Instances of events are called *tasks*. The identifier *TASK* stands for a set of tasks.

Tasks have durations, and their starting and ending points are called *transitions*. For a task x, we use x_s to denote its *start transition*, and x_e for its *end transition*. The identifier *TRANS* denotes a set of transitions.

3.2 Causal Relations of Transitions

There is a binary relation, $<_c$, a *causality relation* on $(TRANS \times TRANS)$, which is an irreflexive partial order; there is also a causal equivalence relation, $=_c$. Intuitively, for transitions w and w', $w<_c w'$ means that w causes w', and $w=_c w'$ means that the two transitions have the same causes and cause the same transitions.

In the set *TASK*, there is a special element, $*$, called *the silent task*, the start of which acts as an "on-button" and the end as an "off-button". In other words, the silent task is present throughout the period in which one is interested.

It seems sensible to stipulate that, for any task x,

$$x_s <_c x_e$$

and for any transition w,

$$*_s <_c w \lor *_s =_c w$$

$$w <_c *_e \lor w =_c *_e$$

Figure 1 presents an example of tasks and causal relations between transitions. Nodes in the figure denote

transitions, and arcs the partial ordering on $(TRANS \times TRANS)$. Note that, if two transitions are causally equal $(=_c)$, they are represented by the same node.

3.3 Assigning Time

Timing is provided in the theory by a partial function, *Time*, which assigns a real number to a transition (of type $TRANS \to \mathcal{R}$). Intuitively, for a transition w, when $Time(w) = r$, we understand that w takes place at the time instance r; when *Time* is undefined at w, i.e., $w \notin \mathbf{dom}Time$, w does not take place at all.

Time is restricted by $<_c$ and $=_c$ as follows; assuming $\{w, w'\} \subseteq \mathbf{dom}Time$,

$$w<_c w' \Rightarrow Time(w) < Time(w')$$

$$w=_c w' \Rightarrow Time(w) = Time(w');$$

further more, we demand that

$$*_s \in \mathbf{dom}Time$$

$$*_e \in \mathbf{dom}Time$$

and for any task x,

$$x_s \in \mathbf{dom}Time \Rightarrow x_e \in \mathbf{dom}Time$$

$$x_e \in \mathbf{dom}Time \Rightarrow x_s \in \mathbf{dom}Time$$

Figure 2 depicts a situation where a few tasks are designated to take place with their start and end transitions assigned time values.

3.4 Temporal Relations

We proceed with definitions of *temporal ordering*, $<_t$, and *temporal equality*, $=_t$, between pairs of a transition and a real number. For transitions w, w' and real numbers r, r',

$$(w,r)<_t(w',r') \Leftrightarrow w<_c w' \land r < r'$$

$$(w,r)=_t(w',r') \Leftrightarrow w=_c w' \land r = r'$$

Based on these relations, causal relations for tasks are defined as follows. Let x and y be tasks with their start and end transitions in the domain of *Time*, and

$$Time(x_s) = r_{xs} \quad Time(x_e) = r_{xe}$$

$$Time(y_s) = r_{ys} \quad Time(y_e) = r_{ye}$$

then

$$x<_i y \Leftrightarrow (x_e, r_{xe})<_t(y_s, r_{ys})$$

$$x<_h y \Leftrightarrow (x_s, r_{xs})<_t(y_s, r_{ys})$$

If $x<_i y$, we say that x is *interior causal* of y; if $x<_h y$, we say that x is *head causal* of y.

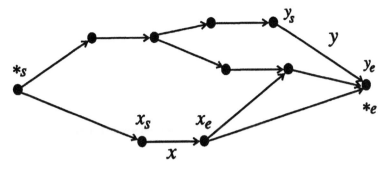

Figure 1: Tasks, transitions and causal relations.

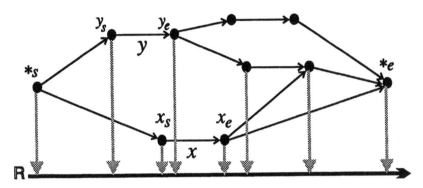

Figure 2: Transitions assigned time.

Figure 3: Interior Causal $<_i$

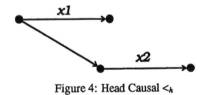

Figure 4: Head Causal $<_h$

Figure 3 and Figure 4 show the two causal relations between tasks. The first one denotes the situation where one task's ending causes another task's starting, whereas the second situation where one task's starting causes another's starting.

3.5 Abbreviations

For convenience, we add some syntactic sugar. We use a predicate *occur* (of type $TASK \rightarrow B$) as an abbreviation such that

$$occur(x) \Leftrightarrow (x_s \in \mathbf{dom}Time)$$

We also define two partial functions, *Start* and *End*, from tasks to reals. For a task x, $Start(x) = r$ if only if $Time(x_s) = r$ (when x_s is not in the domain of *Time*, $Start(x)$ is also undefined); similarly, $End(x) = r$ if only if $Time(x_e) = r$.

3.6 Specification and Proofs

Behaviour of a system is characterized by choice of the following objects in the event model:

- *TASKS*
- $<_c$ and $=_c$
- *Time*

together with their restrictions and properties of real numbers.

To formally reason about the event model as defined above, one needs a *proof theory* for the model. At this stage, we reckon such a proof theory should contain first order logic, a theory of real numbers, and a theory that characterizes the properties above in terms of

3

axioms and inference rules. In the next section, we apply the event model to notions of safety analysis by stating such notions in logical formulae, which should be seen as objects of the proof theory.

4 Interpretation

4.1 Fault Trees

A fault tree can be viewed as a set of events connected by AND-gates and OR-gates. In giving a formal semantics to the fault tree in terms of the event model, we use *tasks* to interpret the events and predicates to interpret the gates.

Because an event in a fault tree may have multiple instances, we need a mechanism to relate such events to tasks in the event model. This mechanism is provided by a labelling function ϕ. In a nutshell, the function ϕ labels the events in a systematic way such that different event instances end up as different tasks.

To ease our argument, we adopt the following naming convention. For a gate with two input events X and Y and the output event Z, we assume that the function ϕ will give us, respectively, tasks which are assigned to variables x, y and z such that $x = \phi(X)$, $y = \phi(Y)$ and $z = \phi(Z)$.

The whole tree is translated into a sequence of predicates, each of which corresponds to a gate; the conjunction of these predicates specifies the tree.

4.1.1 Gates

Here we present a number of schemata or templates for interpretations of gates. For simplicity, we only consider gates with two inputs, as the approach can be easily generalized to handle gates with more inputs.

A function $\mathcal{M}$ is used to interpret the gates of the fault tree; it accepts an intermediate node of the fault tree, *i.e.*, a gate with the input events X and Y, and the output event Z, and returns a boolean expression in the event model — the *characteristic predicate* of the gate.

Causal AND-gate (CAND) For AND-gates involving causality, we give two possibilities.

The first interpretation is a weaker one. Intuitively, an AND-gate means that if the output event Z occurs then both the input events X and Y must have occurred before. To express this in the event model:

$$\mathcal{M}(CAND(X,Y,Z)) \stackrel{\text{def}}{=}$$
$$occur(z) \Rightarrow occur(x) \wedge occur(y)$$
$$\wedge (x <_i z) \wedge (y <_i z)$$

Observe that the "guard" $occur(z)$ is needed because, when z does not occur, the causal relation $<_i$ becomes undefined. Also note that we do not require that z must occur when x and y occur (*i.e.*, occurrences of x and y do not always imply occurrence of z); such a stronger interpretation is also possible, however it is not necessary.

Use of $<_i$ relation indicates that z can only happen after both x and y have finished.

The second interpretation is stronger — if Z occurs then both X and Y must have occurred and the time intervals of X and Y overlap:

$$\mathcal{M}(CAND(X,Y,Z)) \stackrel{\text{def}}{=}$$
$$occur(z) \Rightarrow occur(x) \wedge occur(y)$$
$$\wedge (x <_i z) \wedge (y <_i z) \wedge overlap(x,y)$$

where *overlap* is a predicate accepting a pair of tasks:

$$overlap(x,y) \stackrel{\text{def}}{=}$$
$$\exists r.(Start(x) < r < End(x))$$
$$\wedge (Start(y) < r < End(y))$$

Generalization OR-Gates (GOR) One interpretation, as advocated in [Vesely 81], treats an OR-gate as a generalization in the sense that the output is a restatement of one of the inputs; it is explicitly stated that there is no causality between input and output. This relation can be specified as follows:

$$\mathcal{M}(GOR(X,Y,Z)) \stackrel{\text{def}}{=}$$
$$(occur(z) \Leftrightarrow (occur(x) \vee occur(y))) \wedge$$
$$(occur(x) \wedge \neg occur(y) \Rightarrow z = x) \wedge$$
$$(occur(y) \wedge \neg occur(x) \Rightarrow z = y) \wedge$$
$$(occur(x) \wedge occur(y) \Rightarrow z = x \vee z = y)$$

where = denotes the identity relation between tasks.

Causal OR-gate (COR) Now we come to an OR-gate with causality — the output is caused by one of the inputs.

$$\mathcal{M}(COR(X,Y,Z)) \stackrel{\text{def}}{=}$$
$$occur(z) \Rightarrow (occur(x) \vee occur(y))$$
$$\wedge (occur(x) \wedge \neg occur(y) \Rightarrow (x <_h z))$$
$$\wedge (occur(y) \wedge \neg occur(x) \Rightarrow (y <_h z))$$
$$\wedge (occur(y) \wedge occur(x) \Rightarrow (x <_h z) \vee (y <_h z))$$

Other Properties The above interpretations are just examples to demonstrate the expressiveness of the theory.

Other kinds of gates, such as the EXCLUSIVE-OR gates and generalization AND-gates, can be easily coded as predicates in the model, too; they are omitted here due to lack of space.

An event of fault tree may have timing constraints embedded. For instance, such an event may be stated as *the operator fails to respond to the alarm in 30 seconds*; fault tree analysis lacks mechanism in making use of such timing information. However, the event model allows us to express such properties straightforwardly, and what is more, it allows us to reason about such properties.

4.1.2 Comment

The interpretation of fault tree constructs in the previous sub-sections adds more information to fault trees, and in particular, timing information, which conventional fault tree analysis tends to ignore or to be vague about.

This interpretation brings forth more precision to the user. However, one must realize that precision can give rise to obligations as well as benefits. It is up to the user to decide which gates correspond to which interpretations; more powerful analytical techniques will be needed if one would like to take full advantage of the model; some traditional techniques, such as using the *minimal cut-sets*, may not apply any more, and this point is illustrated as follows.

Obtaining the minimal cut-sets from a fault tree relies on a few laws of boolean algebra, such as one of the distributive laws

$$\alpha \wedge (\beta \vee \gamma) = \alpha \wedge \beta \vee \alpha \wedge \gamma$$

However, if the user chooses to interpret his OR-gates as the causal one (COR in 4.1.1) and his AND-gates as "the overlapping input events cause the output event", that is, the second interpretation of the AND-gates in 4.1.1, the law above no longer holds. The left hand side means that overlapping of α and an intermediate event caused by either β or γ may cause the top event, whereas the right hand side says that either overlapping of α and β or overlapping of α and γ may trigger the top event.

It is worth noting that the distributive law would hold if the user interpreted his AND-gates in a different way, such as the first interpretation in 4.1.1. Nonetheless, the moral is that we must examine the traditional techniques and notions carefully in the light of the event model and be prepared to form new notions and develop new techniques where appropriate.

4.2 Event Trees

To interpret event trees in the event model, we adopt a similar approach as for the fault trees: events are in-

terpreted as tasks by a labelling function ϕ. The basic construct of an event tree is seen as a "branch", namely, an initiating event and two immediately consequent and mutually exclusive events (one of the consequent events represents the success of the initiating event, and the other the failure). We denote such a branch as $BRANCH(X, Y, Z)$ with X as the initiating event and Y and Z as consequent events.

Assuming that the events X, Y and Z correspond to the task variables x, y and z respectively, we use a function $\mathcal{M}$ to give meaning to a branch in terms of a predicate:

$$\mathcal{M}(BRANCH(X, Y, Z)) \stackrel{\text{def}}{=}$$
$$(occur(x) \Leftrightarrow occur(y) \oplus occur(z)) \wedge$$
$$(occur(x) \wedge occur(y) \Rightarrow (x <_h y)) \wedge$$
$$(occur(x) \wedge occur(z) \Rightarrow (x <_h z))$$

where $\oplus$ denotes the exclusive-or operator. In the predicate, the first clause says that X occurs if only if exactly one of Y and Z occurs; the second clause states a causal relation between X and Y, guarded by their occurrences; the third clause is similar to the second, but is for X and Z.

4.3 Common Cause Failures

We demonstrate how a property of the common cause failures can be expressed in the event model.

Given a set of events, Ψ, representing a set of common cause events, and an event X — their common cause. We understand that, if any event Y in Ψ occurs, then all other events in Ψ occur, and further more, X occurred earlier and caused all the events in Ψ. With a labelling function converting events to tasks such that X becomes x, etc., this fact can be expressed by:

$$\forall Z \in \Psi. \, occur(z) \Rightarrow$$
$$(\forall Y \in \Psi. \, occur(y) \wedge occur(x) \wedge (x <_i y))$$

Note that this predicate does not specify whether the events in Ψ should overlap. However, the event model is expressive enough to cover properties like overlapping, and many such subtleties can be distinguished by using different predicates.

5 Concluding Remark

Requirements for precision in our understanding of safety analysis and for a framework to combine applications of different techniques lead us to rely on a formal, or mathematical, approach to safety analysis. This requires a re-examination and possible extension of existing methods with a formal semantics.

The theory of event presented seems a good candidate for such a purpose. It incorporates notions essential to safety analysis, such as causality and timing, in a systematic way; many properties of safety analysis techniques can be expressed easily.

We plan to mechanize the proof theory of the event model and conduct more case studies to evaluate the formalism.

Acknowledgement Partially supported by a grant KBN T/15/060/90-2 of Polish government. The work was undertaken as part of the SEW EUREKA Project EU263.

References

[HSE 87] Health and Safety Executive, *Programmable Electronic Systems in Safety Related Applications, General Technical Guidelines*, Her Majesty's Stationery Office, London, 1987.

[Murphy 90] D Murphy, *Approaching a Real-time Concurrency Theory*, in *Semantics for Concurrency*, edited by M Z Kwiatkowska *et al.*, Springer-Verlag, 1990.

[Vesely 81] W E Vesely, F F Goldberg, N H R ...s and D F Haasl, *Fault Tree Handbook*, United States Nuclear Regulatory Commission, 1981.

THE FORMALIZATION AND ANALYSIS OF
A COMMUNICATIONS PROTOCOL

G. Bruns and S. Anderson

Laboratory for Foundations of Computer Science,
Computer Science Department, University of Edinburgh, Edinburgh EH9 3JZ, UK

Abstract. The MSMIE protocol (Santoline et al., 1989) was designed to allow processors
in a nuclear safety system to communicate efficiently and reliably via shared memory. Our
formalization and analysis shows that the protocol lacks an important liveness property. In
actual operation, timing constraints are checked to avoid potential problems. We present a
modified protocol that possesses the liveness property even without such constraints. We
also show how parts of the analysis were automated with the Concurrency Workbench.

Keywords. Safety; computer software; concurrency; formal methods; temporal logic.

INTRODUCTION

Computer systems involving both hardware and software are increasingly used in safety-critical applications. Guaranteeing reliability depends critically on
the ability to establish properties of computer systems. Here we offer a case study in establishing properties important in assuring the safety of a fragment
of a communication protocol intended for safety-
critical use (Santoline et al., 1989). The Multiprocessor Shared-Memory Information Exchange (MSMIE)
is a protocol developed for use in multiprocessor control systems. The fragment of MSMIE examined
herein fails to have certain properties we consider
relevant to its safe operation. Though the deficiencies we identify are dealt with within the full safety-
critical protocol nonetheless the analysis of the fragment is revealing. The case study exemplifies the use
of Milner's CCS (Milner, 1989) as a means of modelling and reasoning about such systems and the use
of automated tools to aid in the process of analysis.

MODELLING MSMIE

In this section we describe the MSMIE protocol informally and show how it can be formalized in the
notation CCS.

The MSMIE protocol

MSMIE is a communication protocol for use in a distributed system containing some application-oriented
processor boards (called *master processors*), some
single-function processor boards (called *slave processors*), and a shared memory. The master processors,
slave processors, and shared memory are connected
via a bus. Communication between processors is
achieved by having one processor write a data item
to shared memory, and having another processor read
the item from shared memory. In the MSMIE protocol, each shared memory location is dedicated for use
in either master-to-slave communication or in slave-
to-master communication. The master and slave processors operate asynchronously: the slave processor
can write to a memory location at any time, even if
the previously written value was never read by a master processor. Similarly, master processors can read a
memory location at any time, even if the stored data
has not been updated since the previous read.

Two main considerations guided the design of the
MSMIE protocol. The first arises in cases where,
because of its size, a data item must be stored in a
buffer consisting of several shared-memory cells. An
erroneous value can be obtained if a process reads
the item while another process has written some, but
not all, of the cells in the buffer. A shared-memory
protocol must avoid this kind of "data-tearing".

The second consideration is that the protocol should
provide a minimum, deterministic cycle time for communication between processors. Unnecessary waiting
by a reading or writing processor is not acceptable.

Multiple master processors in a system are present
only for the purposes of fault tolerance. Consequently, the MSMIE protocol differs in the details of
master-to-slave and slave-to-master communication.
In this paper we examine only slave-to-master communication. Furthermore, MSMIE has an "access
mode" feature that we ignore.

As a means of explaining the MSMIE protocol, we
will describe two simple protocols and show that they
do not meet the requirements we have just described.

First consider a protocol in which a data item to be
communicated from a slave to a master processor is
stored in a single shared-memory buffer. To avoid

data tearing, a semaphore is used to enforce mutually exclusive access to the memory buffer. Unnecessary delays are caused whenever a slave is forced to wait for a master to finish reading, and conversely.

Next consider an improved protocol that uses two buffers so that the slave processor can write while a master processor reads. We refer to a pair of buffers and a semaphore as a *memory image*. Each buffer holds both a data value and a status value, which coordinates actions of the processors. A buffer has status **slave** while a slave processor is writing data to that buffer. Once the slave has finished writing, it changes the status value to **newest** to indicate that the buffer can be read. A master processor changes a buffer of status **newest** to status **master** before it begins to read the buffer. Once the master has finished reading, it changes the status of the buffer to **idle**, signaling that the buffer is free for use by a slave processor.

A semaphore guarantees mutually exclusive updating of buffer status values. A slave processor may write to a **slave** buffer, and a master processor may read from a **master** buffer, however, without acquiring the semaphore.

Unnecessary delays can still occur with this improved protocol. Consider a state in which there is one **slave** buffer and one **master** buffer. After the slave processor has finished writing, it will change the buffer status to **newest**. The slave processor may be ready to write a new data item immediately, but it must wait for the master processor to finish reading before an idle buffer will become available. The protocol could be modified so that the slave processor would change the **newest** buffer to **slave** if no idle buffer were available, but this would cause the master processor to delay if it was ready to read again while the slave was still writing.

The MSMIE protocol avoids these delays through the use of a third buffer. Each of the three buffers can take the same status values as before. Additionally, each memory image contains a count of the number of currently reading processors. The initial state of one buffer is **slave**, the initial state of the other buffers is **idle**. The number-of-readers count is initialized to zero. The MSMIE protocol is as follows:

Slave write

1. Acquire semaphore

2. If a **newest** buffer is available, change its status to **idle**

3. If a **slave** buffer is found, change its status to **newest**, else an error has occurred

4. If an **idle** buffer is found, change its status to **slave**, else an error has occurred

5. Release semaphore

Master acquire

1. Acquire semaphore

2. If a **master** buffer is not found, then
   if a **newest** buffer is found, change its status to **master**
   else release semaphore and try again

3. Increment number-of-readers count

4. Release semaphore

Master release

1. Acquire semaphore

2. Decrement number-of-readers count

3. If number-of-readers = 0, then
   if a **newest** buffer is found, then
      change status of **master** buffer to **idle**
   else change status of **master** buffer to **newest**

4. Release semaphore

The slave processor executes its part of the protocol after a complete data value has been written to a buffer having status **slave**. The **Slave write** action causes the **slave** buffer to become a **newest** buffer. Some other buffer becomes the new slave buffer that the slave processor will write to next.

The **Master acquire** part of the protocol is performed by a master processor when it is ready to read from a buffer. The result is that some buffer will have status **master**, from which the master processor can subsequently read. Once a master processor has finished reading, it executes the **Master release** part of the protocol, causing the buffer to possibly change status to **newest** or **idle**.

Notice that the actions of the protocol only modify buffer status values. Buffer data values are read and written between protocol actions.

Formalizing the protocol

So far we have described the MSMIE protocol but have not shown that it has the properties we would like. To do so, we must formalize our description of the protocol. We can model the behaviour of the protocol formally using a *behaviour graph*, which is a directed graph in which nodes represent states and labelled edges represent actions.

We will use the notation CCS (Milner, 1989) to provide a convenient syntax for representing behaviour graphs. CCS is sometimes called a *process algebra* because process terms are built up from an inactive process *nil* using a small set of operators. Before describing the CCS process operators, we need to say a little about actions. CCS actions are either simple *names* $(a, b, \ldots)$, *labels* build from names and overbars $(a, \overline{a}, \ldots)$, or *actions* comprising labels and the special synchronization action τ. If two actions differ only because one has an overbar, e.g. a and $\overline{a}$, then we say that they are *complementary* actions.

The semantics of CCS defines the behaviour graph corresponding to every CCS term. A CCS term of the form $\alpha.P$, where α is an action and P a process term, can perform α and thereby evolve to P. We write $\alpha.P \xrightarrow{\alpha} P$ to represent this state transition.

For a term of the form $P + Q$, the transition $P + Q \xrightarrow{\alpha} P'$ is possible if $P \xrightarrow{\alpha} P$ is possible, and the transition $P + Q \xrightarrow{\alpha} Q'$ is possible if $Q \xrightarrow{\alpha} Q'$ is possible.

For a term of the form $P \mid Q$, the transition $P \mid Q \xrightarrow{\alpha} P' \mid Q$ is possible if $P \xrightarrow{\alpha} P'$ is possible, the transition $P \mid Q \xrightarrow{\alpha} P \mid Q'$ is possible if $Q \xrightarrow{\alpha} Q'$ is possible, and the transition $P \mid Q \xrightarrow{\tau} P' \mid Q'$ is possible if $P \xrightarrow{\alpha} P'$ is possible and $Q \xrightarrow{\overline{\alpha}} Q'$ is possible. Note that a τ transition is considered to be the result of internal action by a process, and is therefore invisible to an observer of the process. Thus, a τ transition that arises due to the $\mid$ operator should be regarded as an invisible synchronization action.

For a term of the form $P \backslash L$, the transition $P \backslash L \xrightarrow{\alpha} P' \backslash L$ is possible if $P \xrightarrow{\alpha} P'$ is possible and $\alpha \notin L$.

For a term of the form $P[f]$, the transition $P[f] \xrightarrow{f(\alpha)} P'[f]$ is possible if $P \xrightarrow{\alpha} P'$ is possible. The function f, from actions to actions, is called a *relabelling* function.

The set of observable actions of a CCS process is called its *sort*.

Consider the binary semaphore used to control access to the shared memory buffers. If it is in the available state, then it can perform an *acquire* action to enter the unavailable state. Likewise, from the acquired state it can perform a *release* action to enter the available state. The behaviour graph of a semaphore is therefore as follows:

A CCS description of a semaphore is as follows:

$$Semaphore \stackrel{\text{def}}{=} acquire.release.Semaphore$$

The CCS term on the left-hand-side of the definition is a constant. A constant can perform exactly those actions possible for the process term on the right-hand-side of the definition. We will refer to CCS constants as *agents*. The sort of agent *Semaphore* is $\{acquire, release\}$.

A natural way of describing the MSMIE protocol is to likewise represent the behaviour of the slave processor, the master processor, and the shared memory as CCS agents. A CCS description of the behaviour of a slave processor is as follows:

$$Slave \stackrel{\text{def}}{=} \overline{acquire}.Slave_1$$

$$Slave_1 \stackrel{\text{def}}{=} \sum_{j=1}^{3} is\_newest_j.\overline{to\_idle_j}.Slave_2 + no\_newest.Slave_2$$

$$Slave_2 \stackrel{\text{def}}{=} \sum_{j=1}^{3} is\_slave_j.\overline{to\_newest_j}.Slave_3$$

$$Slave_3 \stackrel{\text{def}}{=} \sum_{j=1}^{3} is\_idle_j.\overline{to\_slave_j}.Slave_4$$

$$Slave_4 \stackrel{\text{def}}{=} slave.\overline{release}.Slave$$

Note the close correspondence between this formal description and the informal description given earlier. The *slave* action in the definition of $Slave_4$ occurs whenever the slave has completed its part of the protocol. The need for such an action will become clear in the analysis of the protocol. The notation $\sum_{i=1}^{n} P_i$ is just shorthand for $P_1 + \ldots + P_n$.

The behaviour of a master processor is described as a CCS agent *Master* having a sort that includes m_acq and m_rel. The action m_acq occurs when a buffer is acquired for reading; the action m_rel occurs when a buffer is released after reading. We will include two master processors in our model of the MSMIE protocol. The CCS relabelling operator can be used to define the two copies we need:

$$Master_i \stackrel{\text{def}}{=}$$
$$Master[m\_acq_i/m\_acq, m\_rel_i/m\_rel]$$

The behaviour of a set of three memory buffers and the number-of-readers count is described as a parameterized CCS agent $Buffers(b_1, b_2, b_3, r)$. The parameters b_1, b_2, and b_3 range over buffer status values. The variable r represents the set of reading master processors, and ranges over $2^{\{1,2\}}$, where 1 and 2 identify the master processors. For each state s in $\{idle, newest, slave, master\}$, and for each j from 1 to 3, the sort of *Buffers* includes an action of the form is_s_j, which can occur if the jth buffer is in state s, and an action of the form to_s_j, which can always occur and causes the jth buffer to have state s. The sort of *Buffers* also includes the actions no_newest and no_master, which can occur if no buffer has state **newest** or **master**, respectively, and the actions $zero$, inc, and dec, which manipulate the set of reading agents.

We do not model the buffer data values, only the status values, since the data values do not play a part in the behaviour of the protocol.

Composing the components with the CCS operator $\mid$, we get the agent *Msmie*, which describes the behaviour of the protocol:

$$Msmie \stackrel{\text{def}}{=}$$
$$(Semaphore \mid Slave \mid Master_1 \mid Master_2 \mid$$
$$Buffers(slave, idle, idle, \emptyset)) \backslash L$$

The action set L restricts all actions related to communication between components, leaving the set $\{slave, m\_acq_1, m\_acq_2, m\_rel_1, \text{ and } m\_rel_2\}$ as the sort of *Msmie*.

REASONING ABOUT MSMIE

Having a precise model of a fragment of the MSMIE protocol allows us to reason about it. Though there is no explicit specification to which MSMIE must conform, there are some properties which such a protocol must satisfy and some which, though not essential, might influence the choice of one protocol over another. In this section some of these properties are introduced and the MSMIE protocol is found to lack a property we might reasonably expect it to possess.

The statement of the properties discussed in this and the next section is entirely informal. The main reason for this is to present the results of this paper in an easily accessible manner. All of the properties used here can be formalized in the modal mu-calculus (Stirling, 1989). Using the Concurrency Workbench (Cleaveland et al., 1989) it is possible to check automatically whether such a formal property is satisfied by a CCS agent. Here, when a property is claimed to hold (or fail to hold) for a protocol, the claim has been checked using the Concurrency Workbench.

A simple property which should be satisfied by any protocol is deadlock freedom, which guarantees that the protocol is always capable of responding in some way. A CCS agent is deadlock free if it is always capable of carrying out an action. The MSMIE protocol is deadlock free.

Similarly, the MSMIE protocol possesses the property that a slave action is always possible. In (Santoline et al., 1989) this property is cited as a critical factor in the design of the protocol. This means that a slave processor will never be delayed from writing a buffer by the protocol.

We might want to check that master processors get fresh information reasonably often. We call a buffer *fresh* if its status is newest and its previous status was slave. Not every buffer with status newest is fresh; such a buffer may have previously had status master. To check that master processors get fresh information it is necessary to augment the protocol to include a new action: *update*. *Update* actions are inserted into the protocol to indicate that on the next m_acq_k action a fresh buffer will be acquired by the kth master. This will be the case when either:

- a *slave* action occurs and there is no buffer with status master or,

- an m_rel_k action occurs, there is a buffer with status newest, and k is the only master currently reading a buffer.

The sole reason for the introduction of *update* actions is to aid in the analysis of the protocol. By analogy with the earlier property of *slave* actions it might be reasonable to expect that it is always the case that an *update* action eventually happens. The MSMIE protocol fails to have this property, in part because there is no presumption in the description that each of the processors make progress. Thus, there are behaviours in which the slave processor ceases to supply fresh information and therefore after the last information supplied by the slave has been read by the master processors no further *update* action can occur.

A further refinement of the property concerning *update* actions is to restrict attention only to behaviours in which all the processors are active. We thus consider the property that if the master and slave processors all contribute an infinite number of actions to the behaviour of the system then an infinite number of *update* actions occur in the behaviour.

But, does this property hold of the MSMIE protocol? No, at least not for our CCS model of the protocol, due to traces in which a buffer remains in the master state indefinitely. For example, consider the *Msmie* state in which the buffer status values are slave, master, and newest, and in which the master processor identified by 1 is reading a buffer. If the actions *slave*, $m\_acq_2, m\_rel_1, slave, m\_acq_1, m\_rel_2$ occur in sequence, then *Msmie* will be return to the same state. *Msmie* could repeat this same sequence of actions indefinitely without an *update* action ever occurring.

IMPROVING MSMIE

In the previous section a potential flaw in the MSMIE protocol was identified in the presence of multiple master processors. The possibility of such a flaw is recognised in (Santoline et al., 1989) and timing constraints are used in the presence of multiple masters to avoid the "lockout" described in the previous section. Though this approach may be entirely satisfactory in practice the question remains: is it possible to modify the MSMIE protocol so that additional timing constraints are unnecessary in the presence of multiple masters? In this section a modified MSMIE protocol which avoids "lockout" is presented and analysed.

The improved protocol

The modified protocol adds to MSMIE a fourth buffer and an additional buffer status: old, standing for *old master*. In the new protocol, a master processor may be reading from either the master or the old master buffer, and so the parameter r of agent *Buffer* is replaced by two parameters: r_m and r_o. Readers are *never* added to those reading from the old master buffer; thus eventually it will become free (i.e. have zero readers) provided all its readers make progress. In the new protocol, when a master processor, k, carries out a m_acq_k action there are 3 cases to consider:

- If there is no buffer with status master the buffer with status newest is relabelled master and k is the reader of the buffer.

- If there is a buffer b with status master, a buffer b' with status newest and no buffer with status old, then b is relabelled old (retaining its readers) and b' is relabelled master and k is its reader.

- If there is a buffer b with status master and a buffer with status old, then k is added to the set of readers of b.

Provided all the masters and the slave make progress the old master always becomes free eventually (because its set of readers is never increased) and fresh data supplied by the slave will be read.

Analysis of the improved protocol

In order to state precisely the property we want we must add *update* actions to the modified protocol. *Update* actions are inserted at those points where we know that the next m_acq_k will commence reading a fresh buffer.

The property which should hold of the modified protocol is:

> *For any behaviour of the protocol beginning at its initial state, if slave and m_acq_k actions happen infinitely often then update happens infinitely often.*

Using the Concurrency Workbench it is possible to mechanically check that the property does hold for this modified version of the protocol with two master processors. However it is instructive to sketch an informal proof that the property holds for the protocol. The proof sketch depends on two observations which are easy to prove:

- If the state of the protocol is such that there is a buffer labelled **newest** and both r_o and r_m are non-empty, then as long as the r_o and r_m components remain non-empty there will be a buffer labelled **newest**.

- For any state S of the protocol and for any behaviour starting from S, if m_acq_k appears infinitely often for each master processor k, then the protocol passes through a state in which the r_o component is empty.

The proof sketch is by contradiction. If the property fails to hold there must be some state S of the protocol and behaviour beginning at S in which *slave* and m_acq_k actions happen infinitely often but *update* never happens. Suppose this is true. By considering the protocol we can see that *slave* actions can only occur when the r_o and r_m components of the state are both non-empty (otherwise an *update* occurs). So at some point we arrive at a state where we have a buffer labelled **newest** and both the r_o and r_m components of the state are non-empty. From the two observations above we can see that eventually we carry out an m_rel_k action which results in the r_o component becoming empty when there is a buffer labelled **newest**; by inspecting the protocol we can see that an *update* must occur. This contradicts the assumption that an *update* does not occur.

AUTOMATED ANALYSIS

Much of the reasoning used in the analysis of the MSMIE protocol can be encoded into algorithms and thereby mechanized. The Concurrency Workbench (Cleaveland et al., 1989) (which we will refer to simply as the *Workbench*) is a software tool for the analysis of finite-state CCS expressions. The two most important forms of analysis it can perform are equiv-

alence checking and model checking. Equivalence checking can be used to show that a specification and an implementation, both expressed as CCS expressions, exhibit the same behaviour. Several kinds of equivalence relations are supported by the Workbench, and preorder relations between processes can also be checked. Model checking can be used to show that a property holds of a process.

Other features of the Workbench include a minimization function that generates a minimal CCS expression (in number of states) that is observation equivalent to a given expression, and a synthesis algorithm that solves equations of the form $(A|X) \setminus L = B$, where CCS expressions A and B, and action set L are given. The Workbench can also perform more mundane tasks such as computing all successor states possible of a given expression.

It is important to understand that most features of the Workbench require that agents be finite-state and written in basic CCS. Since our definition of the *Msmie* agent included some features of value-passing CCS, such as parameterized agents, we could not analyze the protocol directly with the Workbench. This problem led to the development of a a language for value-passing CCS based on the definition given in Chapter 2 of (Milner, 1989), and a translator from this language to basic CCS. After translating our value-passing agent definitions to basic CCS, we were able to analyze the protocol with the Workbench.

CONCLUSIONS

A careful analysis of the MSMIE protocol was made possible by formalizing it in the notation CCS. The theory of CCS provides a sense of behavioural equivalence that was used in the analysis. The modal mu-calculus was used to formalize some desirable properties of the protocol, which led to the discovery that the MSMIE protocol lacks an important liveness property. We developed a modified version of the protocol that has the property. By formalizing the protocol, we also made possible the automated analysis of the protocol with the Concurrency Workbench.

As already mentioned, the flaw we have shown in the original protocol does *not* cause problems in real applications, due to the use of timing constraints. Our modified protocol behaves properly without the use of timing constraints.

Some notable problems arose during the course of our analysis:

- It was difficult to express some properties of interest in the modal mu-calculus.

- The Concurrency Workbench provided too little information as output from the equivalence checking and model checking commands.

- The mechanical analysis only gave a result for the case of a MSMIE system with two masters and one slave processor; the results from the hand proof were more general.

One solution to the first problem would be to provide another temporal logic for use with the Concurrency Workbench, although other popular logics are less expressive than the modal mu-calculus. The second problem could be solved, in the case of equivalence checking, by showing a distinguishing proposition in the case that two expressions were not equivalent. Such an approach is taken in other analysis tools (Godskesen et al., 1989). The third problem suggests future research; some results are described in (Clarke and Grumberg, 1987).

REFERENCES

Clarke, E. and Grumberg, O. (1987). Avoiding the state explosion problem in temporal logic model checking algorithms. In *Proceedings of the 5th Annual ACM Symposium on Principles of Distributed Computing.*

Cleaveland, R., Parrow, J., and Steffen, B. (1989). The concurrency workbench: A semantics based tool for the verification of concurrent systems. Technical Report ECS-LFCS-89-83, Laboratory for Foundations of Computer Science, University of Edinburgh.

Godskesen, J. C., Larsen, K. G., and Zeeberg, M. (1989). TAV users manual. Technical Report R 89-19, Institute for Electronic Systems, University of Aalborg, Denmark.

Milner, R. (1989). *Communication and Concurrency.* Prentice Hall International.

Santoline, L. L., Bowers, M. D., Crew, A. W., Roslund, C. J., and III, W. D. G. (1989). Multiprocessor shared-memory information exchange. *IEEE Transactions on Nuclear Science*, 36(1).

Stirling, C. (1989). *Temporal Logics for CCS*, volume 354 of *Lecture Notes in Computer Science*, pages 660–675. Springer Verlag.

FORMAL CONSTRUCTION OF HARD TIME REQUIREMENTS IN THE SOFTWARE DESIGN

R. Bareiß

*Institute for Control Engineering and Industrial Automation, University of Stuttgart,
Pfaffenwaldring 47, W-7000 Stuttgart 80, Germany*

Abstract:

The necessary supposition for an error and disturbance free interaction of the process-control system with the technical process is the guaranteed meeting of any possible time constraints, which are based on the necessities of the technical process. Keeping this aim in mind, a contribution for attaining it, by ways of formal verification, is presented in this paper. The approach allows to assure the correctness of specified quantitative temporal constraints among the different formal parts of the specification created in the various stages of the software development process.

Keywords: *Verification, time state machines, safety, embedded real-time systems.*

1. Introduction

In the field of process automation the used process-control systems in general have to fulfil hard real-time constraints. It is not an untypical situation that a real-time control system has to control several parallel and time critical processes simultaneously. The necessary supposition for an error free and disturbance free interaction of the reactive process-control system with the technical process, is the guaranteed meeting of the time requirements based on necessities of the technical process (see, e.g. [JaLe89])

This may be a safety relevant attribute. We just have to think of an anti-block system in a car. The processor always has to react at the right time, otherwise the breaks wouldn't work in the desired way. The necessity of meeting hard time requirements can also be realized by inspecting a lift with, for instance, three cages. In the worst case every cage is moving. And being a comfortable lift, the speed of the cages is controlled between the floors. While usually the behaviour of the lift is managed by one computer, it is part of three closed control loops in this situation, working simultaneously. Thus, very hard time constraints have to be met resulting from the cyclic sampling of the speed information, and resulting from the calculation of the associated output values of the simultaneously working control loops. Violating the time constraints, caused e.g. by over-load of the computer, may lead to a wrong cage speed or worse, to a instable closed control loop with its dangerous consequences to the plant or even to human being. Because of its being safety relevant in most real-time systems, there is the necessity of guaranteeing the meeting of external time constraints.

Aiming to contribute to the solution there already exist some approaches each enlightening different aspects. For instance, one of them is based on efficient scheduling algorithm, like deadline scheduling and relating algorithms as discussed for example in [Roes79] or [RSS90]. These algorithms are quite useful because of their combining an increased probability of meeting time requirements with an increase of work-load compared to real-time applications based on priority driven scheduling concepts. But giving a guarantee of meeting hard time requirements is not possible.

Another approach based on schedulability analysis (see e.g. [Stoy 87/1], [Stoy 87/2], [Agne91]) promises to assure the meeting of time requirements specified formally on a very low and implementation-close level by an implementation on a certain computer. But how far can we trust that detailed software design being error free, particularly with respect to time requirements? This is a central issue especially in large real-time systems. It's not enough just to pay very much attention to avoid errors during the development process. In our opinion, more work is necessary. It has to be assured formally, that the time requirements in the software design are correct against external time constraints.

With our approach we're sure to reach this aim. Starting from a formal description of the behaviour of the technical process showed at the interface, the step-wise development of the associated controlling part is aided. At the end a detailed software design fulfilling the external time constraints will exist.

As a formal specification method we use a method based on the concept of communicating state machines. It is a very powerful method especially for formal requirements analysis (see, e.g. [JLHM90]) and leads to a specification which is very easy to survey.

According to the character of the specification method, the application of formal methods as presented in this paper is intended to support the development process in a constructive manner. It has to provide a correct basis for the succeeding development steps always. And not by applying a verification afterwards to proof that something doesn't happen inside our specification seen as a black box.

The organization of the paper is as follows: Section 2 presents an overview of the development modell for a real-time system. This is the basis for the further considerations. In section 3 the specification method is discussed and presented. An small but complete example is introduced in order to illustrate the formal construction process of time requirements. And section 5 describes the procedure for assuring the desired correctness.

13

2. Development Model for Real-Time Systems with Hard Safety Related Time Constraints

Figure 1 shows the software development process model for real-time systems.

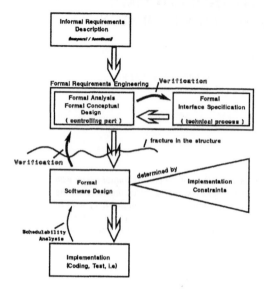

Figure 1: Software development process model for real-time systems

The two major activities in the software development process are the formal requirement analysis and the development of the formal software design. In the formal requirement analysis phase, first of all, the attention has to be focused on the formal description of the behaviour of the technical process showed at the interface. That means:

- Describing the communication paths between the technical process and the controller together with their characteristics. For instance, defining the data interchange and the existing interrupts.
- Specifying the quantitative temporal requirements regarding the communication paths.
- Describing the technical processes' internal temporal and functional behaviour, as far as it's necessary for the cooperation of the control computer with the technical process.

Afterwards, the behaviour of the controller has to be specified formally. This specification is determined by the functional and temporal requirements for the real-time system fixed already earlier informally, and the formally specified behaviour of the technical process. (A temporal requirement, for instance, may be: The controller has to be able to control three closed control loops simultaneously and further has to serve at least a certain interrupt four times a second). The so called conceptual design could be characterized by the following pecularities:

- It is a formal description of the general, conceptual solution including the functional and complete temporal aspects of the controller without paying respect to its realisation.
- The structure is only determined by the problem.

If an abstract machine existed for running our formally specified conceptual solution, the real-time system would be complete. But because of having a computer as a basis for realisation, it has to be specified in a subsequent phase how to implement the conceptual design on a concrete control computer with a concrete operating system and a concrete target language in mind. This leads to the software design, including the following features:

- Specification of parallel tasks (possibly with their priority), and specification of synchronization and communication relations between the tasks.
- Identification and definition of procedures.
- Specification of temporal requirements associated with the communication and synchronisation relations.
- The structure is further determined by a lot of functional and non-functional external constraints, namely modularization criteria of the software, limited storage space, decoupling the i/o-handling from the processing, etc.

Therefore, in practical work the phenomenon occurs that the structure of the software design differs largely from the structure of the conceptual design (mentioned in figure 1 as 'fracture in the structure'). That is why the creation of the software design is up to now a process which is demanding creativity and human intelligence.

In the last stage, the source code itself may be developed or partially created automatically on the basis of the now existing detailed software design.

Aiming primarily to support the work of the engineer in the most critical phases and therefore most necessarily support needing phases, namely the formal requirements analysis phase and the following formal software design phase, our attention is first of all focused on the following verification steps:

- First, it has to be assured, that the specification of the temporal aspects in the conceptual desing fulfils the required hard time constraints of the technical process.
- Second, it has to be assured, that the specified temporal requirements in the conceptual design will be fulfilled by the temporal requirements in the software design.

Therefore, in the further parts of this paper, the specification concept underlying these two main stages will be discussed to further detail and afterwards the formal verification activity will be illustrated on a small, but complete example. The presented concept supports the interactive construction of time reuirements, based on the structure of the associated specification and the quantitative temporal constraints of the previous specification phase.

3. Our Specification Method

In order to realize an even for practitioners easily usable prototype of a formal specification and verification tool, the choice of the concepts for the specification method has to be done very carefully. A concept based on communicating state processes seems ideal because of the following reasons:

- The method has proved to be very applicable in practical use (see, e.g. [HLNP90]). It is a problem close and solution oriented specification method.
- Especially for the requirements analysis this method leads even for large systems to an easy to survey and easy to understand specification.
- It is a well known and widespread method (see, e.g. [MeWa86], [Ostr89], [OsWo90], [HaPi87]).
- The basic ideas are very simple and easy to understand.
- This method integrates the specification of functional and temporal aspects.

But, however, for providing formal verification with respect to temporal aspects, some very essential extensions become necessary additionally, regarding the underlying semantics and regarding the capabilities of formulating quantitative temporal requirements. This has been done by us. The formalisation and enhancement of the basic concept led in its full extent to a proper edifice of thoughts. The essential parts of it will be presented in the following.

3.1 Proper Definition of the Basic Structure

A state process consists of a set of states (s_i) and a set of transitions (t_i). For unique identification each state process has got his own, user defined, identifier.

The internal behaviour (the control flow) of a state process is defined by a net of states connected with transitions. An idea of how to the use the states and transitions may be given by the following simple example:

- The process 'elevator car' has got the states,
  - s_1 = 'wait',
  - s_2 = 'move up',
  - and s_3 = 'move down',
- and it has got the transitions:
  - $t_1 = s_1 \rightarrow s_2$ ('wait' -> 'move up')
  - $t_2 = s_2 \rightarrow s_1$ ('move up' -> 'wait')
  - $t_3 = s_1 \rightarrow s_3$ ('wait' -> 'move down')
  - $t_4 = s_3 \rightarrow s_1$ ('move down' -> 'wait')

Considering only the temporal aspects, a state process remains in a state for a certain time while it takes an infinite short time for a transition to occur. Formally:

- $\text{duration}(t_n) = 0$
- $\text{duration}(s_n) = a$ and $a >= 0$

Taking now the functional aspects into consideration we realize two kinds of states. One type of state is a pure 'waiting'-state as the state s1, where no information processing takes place. We assume with respect to a software system that no cpu-power is used. And the other state is something like a 'dynamic' state, in which information processing occurs. I.e. the movement of the cage over time. Relating to a software system this means that during remaining in a 'dynamic' state, a procedure or a function needing cpu-power is processed. In the specification language, this fact is taken into account by providing the possibility of associating the process states with procedures or functions.

The concept presented allows us to have a twofold view on the specification. An abstract one, in which we only see transitions and states (perhaps additionally with the attribute 'dynamic' or 'waiting'). And a concrete one, which shows us especially in the software design the procedures and parallel tasks to be implemented.

3.2 Communication and Synchronisation

A whole specification consists of several state processes, communicating with each other by using a channel with exactly defined behaviour. While the information processing itself takes place while being in a state, a communication process is handled by a transition.

A communication process starts with a sending event which is initiated by the associated transition in the sending process. The sending event for its part initiates the assignment of a sending process information (the content of a local variable) to the channel.

A receiving event may occur, when the receiving process realizes that the channel is not empty any longer and if the other conditions additionally influencing the occuring of the associated transition in the receiving process will be fulfilled too. In that case, the channel content is assigned to a receiving process local variable and the channel is empty again afterwards. But, however, if the associated transition isn't able to occur at that moment, the sent information is stored until the other preconditions for making that transition happen are fulfilled. Accordingly, a channel combines the behaviour of a storage for one value with a proper access concept.

Note, this interesting communication concept is only one of several concepts which are part of the language definition. But it's the most powerful and flexible concept and allows the specification of synchrone and asynchrone communication between parallel processes in a very convenient way.

Because of lacking space, the other concepts are not presented here.

Considering only the temporal aspects, a channel defines a directed sequence relation between at least two parallel processes. Formally defined:

- time-point(sending) < = time-point(receiving)

Synchronisation between parallel processes can be achieved by linking several transitions each of them in a separate process with an event label. These transitions are now coupled to occur at one time, if the conditions for occuring in each state process are fulfilled.

Generally, the conditions to be fulfilled to make a transition happen are:

- A transition may only occur, if the state process is in the state before the appropriate transition.
- The optional specifiable logical condition (a predicate) linked to the appropriate transition has to be true. (This concept allows branching inside of the state processes).
- All receiving channels linked with the appropriate transition have to be filled up with data. That means, all correspondig sending events must already have happened.
- The conditions for the occuring of the other transitions, switched in parallel with the viewed one by an event label, have to be fulfilled too.
- Meeting the specified quantitative temporal requirements (see next section).

3.3 Quantitative temporal requirements

Up to now, we're able to specify the complete temporal behaviour of a real-time system on a qualitative level in a very convenient and vivid way. But for verifying the correctness of quantitative temporal requirements, additional language concepts are needed.

As already mentioned as a demand for quantitative temporal requirement specification in [Laub89], the concepts implemented in the developed formal method capture the specifiability of absolute and relative temporal requirements. Therefore, facilities like the following exist:

- Example for specifying an absolute and cyclic temporal requirement
  - ABS-TIME: $\text{time}(t_n) = k * 100 +/- 5$
    with $k = 0..\text{infinite}$

  This means: The transition t_n occurs inside an interval of $+/- 5$ time units (may be milliseconds) all 100 time units with k running from 0 to an upper limit of 'infinite'.

- Specifying an absolute unique temporal requirement
  - ABS-TIME: $\text{time}(t_n) = 10 +/- 1$

  (The transition tn occurs between 9 and 11 $(10 +/- 1)$ time units after system start.)

- Specifying a relative temporal requirement
  - REL-TIME: $\text{time}(t_n) = 10 +/- 2$

  (The transition time t_n occurs after being in the process state before the accompanying transition tn for a temporal interval between 8 and 12 time units.)

4. Specification of the Example for Illustrating the Verification Process

As being needed very often in the field of process automation and always causing difficulties because of its hard time requirements resulting from cyclic scanning, the closed control loop, where the computer takes the part of the controlling means, seems to be an interesting example especially from a practical point of view.

For this chosen example, a sampling rate of 100 milli seconds will be assumed with a tolerance interval of $+/-10$ ms (in order of better visualising the temporal behaviour in diagrams).

The conceptual design on the level of process communication is visualised in Figure 2. The processes 'technical-process' as the controlled system communicates with the process 'controlling-means' over the channels 'input' and 'output' (from the controlling means viewpoint).

Figure 3 shows the accompanied state charts. Each specified state process consists of one state and one transition passing from the single state to itself. They are initiating the also sketched communication over the channels. In order to enforce the desired simultaneous data exchange, both transitions are linked by the event label 'alpha'. Thus, we could now make the simple conclusion that the transition t_2 occurs every 100 +/- 10 ms, too. Referring to the process development model in Figure 1, the first verification step is trivial in this example.

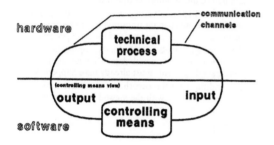

Figure 2: Dataflow of the closed control loop on the level of the conceptual design

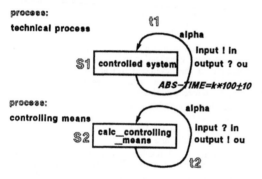

Figure 3: Conceptual design of the closed control loop described with state processes

The temporal behaviour of the complete real-time system on the level of conceptual design, including the timed communication, is visualised in Figure 4. The hatched parts are symbolizing the tolerance intervals.

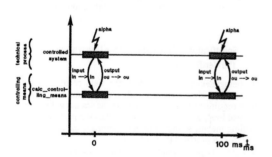

Figure 4: Temporal behaviour of the complete real-time system on the level of the conceptual design.

We now possess a very smart and 'problem close' conceptual design in which we have specified formally and clearly the temporal behaviour including the communication. And we also have defined formally the structure of the problem solution relating to the functional parts by referencing the procedure 'calc_contr_means_val(in,ou)'. But in spite of this advantages, the specification isn't suitable as a direct basis for implementation, because of not managing the calculation of the output values by the procedure 'calc_contr_means_val(in,ou)' in nearly exact 100 ms on a process computer with several other tasks in parallel.

The question of how to implement the already fixed temporal and functional behaviour is part of the software design. It was created here by dividing the process 'controlling_means' into the processes 'handle-input', 'calculating' and 'handle-output', which could be later on implemented as singular tasks. Additionally something like a clock for cyclic activating the i/o-tasks is needed, because in reality the technical process doesn't control the computer by sending the data actively, but the computer, however, has to pick up and send the data from or to the technical process.

The developed software design is illustrated in Figure 5 and 6. Figure 5 shows the communication between the state processes and Figure 6 visualises the internal behaviour of the state processes.

In the initial state, each state process is in the 'wait'-state. Then, initiated by the process 'clock', we will assume that every 100 ms - 10 ms (90 ms, 190 ms, 290 ms, etc.) (it is an explizit design decission) the processes 'handle-input' and 'handle-output' are passing into the accompanying dynamic states in order to pick up the input data from the 'technical process' and sending it to the process 'calculation' or in order to receive the output data from the process 'calculation' and sending it to the 'technical process'. The communication is finished with the according transitions back to the process state 'wait'. After transition $t_{3,2}$, the transition $t_{4,1}$ can occur and the calculation of the output value on the basis of the received input value can start.

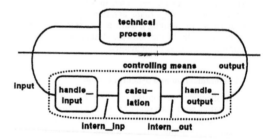

Figure 5: Dataflow of the closed control loop on the level of the software design

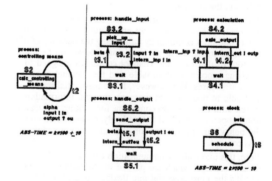

Figure 6: Software design of the closed control loop described with state processes

5. Proving Correctness with Respect to Quantitative Temporal Requirements between the Conceptual Design and Software Design

The point of departure is now a conceptual design with correct qualitative and quantitative temporal requirements and a software design being uncomplete regarding quantitative temporal statements. Desireable, however, would be a software design including all quantitative temporal requirements and, of course, being correct against the temporal requirements in the conceptual design. In order to achieve this for a user in a most convenient and easily understandable way, the following procedure has been developed. It supports formal construction of the missing time requirements. The underlying strategy leads to a worst case estimation with the statement that if the derived quantitative temporal intervals are met by the program, the required temporal time constraints will be met to. The procedure consists of three steps.

5.1 Step 1: Deriving an equality system

In a first step, all qualitative temporal relations of software design will be extracted and mapped into a net by using the following rule:

- All transitions occuring at a time point will be mapped into the nodes
- All states as well as communication relations are transformed in a directed edge between two nodes. This illustrates the predecessor successor relation between two transitions based on the remaining in a state for a certain time or having a time gap between the sending and receiving event.
- All synchronisations are transformed in a not directed edge to illustrate, that there exist several transitions occuring at one time.

In the concrete example this leads to the net as shown in Figure 7 without the parts with broken lines.

Afterwards, the same will be done with the conceptual design and then both nets are put together in a way, that the temporal equivalence of the communication with the technical process in both specifications is guaranteed (Figure 7). These links become - as presented later - the verification conditions.

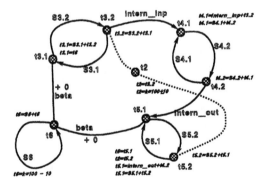

Figure 7: Net generated by extracting the temporal relations from the software and conceptual design

This statement had been already fixed during the creation of the software design. It is based on a explizit design decision.

The assurance of the correctness referring to time requirements in the software design means the solving of this equality system. I.e. the calculation of all intervals regarding the durations of staying and all intervals with respect to the possible time points of transitioning.

Based on this net, two types of equals could be derived:

Type 1: Describing the predecessor successor relation in a way as:

- t_p-of-occurence$(t_{3.1})$ = duration-of-stay-in$(s_{3.1})$ $+ t_p$-of-occurence$(t_{3.2})$

  (t_p = time point); or written shorter:

  $t_{3.1} = s_{3.1} + t_{3.2}$

Type 2: Describing the fact that two transitions happen at the same time

- $t_{3.1} = t_6$

Applicated to the whole net, the following equality system could be derived (see Figure 7 too):

- $t_{3.1} = t_6$ (1)
- $t_{3.1} = s_{3.1} + t_{3.2}$ (2)
- $t_{3.2} = s_{3.2} + t_{3.1}$ (3)
- $t_{4.1} = intern\_inp + t_{3.2}$ (4)
- $t_{4.1} = s_{4.1} + t_{4.2}$ (5)
- $t_{4.2} = s_{4.2} + t_{4.1}$ (6)
- $t_{5.1} = intern\_out + t_{4.2}$ (7)
- $t_{5.1} = s_{5.1} + t_{5.2}$ (8)
- $t_{5.1} = t_6$ (9)
- $t_{5.2} = s_{5.2} + t_{5.1}$ (10)
- $t_6 = s_6 + t_6$ (11)
- $t_{3.2} = t_2$ (12)
- $t_{5.2} = t_2$ (13)

Equals (1) to (11) are describing the temporal relations of the software design.
Equals (12) and (13) are the verification condition.

As further information for solving this equality system we already know the interval in which occuring the transition t_2 occurs, the cyclic sampling based on:

- $t_2(k) = k*100 +/- 10$ (14)

And we know the transition t_6 occuring in the process 'clock' with

- $t_6(k) = k*100 - 10.$ (15)

This statement had been already fixed during the creation of the software design. It is based on a explizit design decision.

The assurance of the correctness referring to time requirements in the software design means the solving of this equality system. I.e. the calculation of all intervals regarding the durations of staying and all intervals with respect to the possible time points of transitioning.

5.2 Step 2: The causality analysis

In the causality analysis the transitions are assigned to the intervals of the system determining and wholly underlying time scale. The meaning of this can be easily illustrated by inspecting equal (2) and (3). Equal (2) says $t_{3.1} = s_{3.1} + t_{3.2}$ and equal (3) says $t_{3.2} = s_{3.2} + t_{3.1}$.

If we consider the presented equals without the systems context, the solution of $s_{3.1}$ and $s_{3.2}$ must be 0. But this is no solution within our interpretation. Therefore, the transitions had to be related to the intervals of the system underlying cyclic time scale. This may lead to the equals (2') as $t_{3.1}(k+1) = s_{3.1} + t_{3.2}(k)$

and (3') as $t_{3.2}(k) = s_{3.2} + t_{3.1}(k).$

For $s_{3.1}$ and $s_{3.2}$ values greater than zero can be then calculated. The same effect can be observed by looking at equal (11). s_6 is greater than zero if (11) is enhanced to

$t_6(k+1) = s_6 + t_6(k).$

The applying of the causality analysis to the upper equality system leads to:

- $t_{3.1}(k) \quad = t_6(k)$ (16)
- $t_{3.1}(k+1) = s_{3.1} + t_{3.2}(k)$ (17)

  temporal relations of

- $t_{3.2}(k) \quad = s_{3.2} + t_{3.1}(k)$ (18)

  the software design

- $t_{4.1}(k) \quad = \text{intern\_inp} + t_{3.2}(k)$ (19)
- $t_{4.1}(k+1) = s_{4.1} + t_{4.2}(k)$ (20)
- $t_{4.2}(k) \quad = s_{4.2} + t_{4.1}(k)$ (21)
- $t_{5.1}(k) \quad = \text{intern\_out} + t_{4.2}(k)$ (22)
- $t_{5.1}(k+1) = s_{5.1} + t_{5.2}(k)$ (23)
- $t_{5.1}(k) \quad = t_6(k+1)$ (24)
- $t_{5.2}(k) \quad = s_{5.2} + t_{5.1}(k)$ (25)
- $t_6(k+1) \quad = s_6 + t_6(k)$ (26)
- $t_{3.2}(k) \quad = t_2(k)$ (27)

  verification conditions

- $t_{5.2}(k) \quad = t_2(k+1)$ (28)

The causality analysis could be fully automated. The algorithm detects if there is any cycle at all and if so, the relations of transitions to the certain intervals will be made automatically.

After the successful application of the causality analysis, the engineer additionally has the full certainty of having a lifelock and deadlock free software design.

5.3 Step 3: Solving the equality system

The formalization of solving the resulting equality system leads to three problem domains, which had to be taken into account:

- The equality system itself has got the following characteristics:
  - two types of equals as represented for instance by equal (16) and (17)
  - the equals are linear dependent
  - each variable represents an interval
  - the number of equals is not fixed.
- There's no statement possible in advance which variables of the equality system are generally known before starting the solution of the equality system, nor is there any statement possible in advance, if the set of known variables is sufficient for the solution of all equals. Therefore, the solution algorithm had to be flexible enough to solve those parts of the equality system automatically which can be solved with consideration of the known values. The remaining part then had to be extracted and presented to the user in an easy to survey way, in order to ask him to give more information. Then the user has to make a further design decision. Afterwards, the algorithm will be started again. This procedure has to be repeated until the equality system is solved.
- Because of each variable not only representing one value but an interval with two values, the equality system is given a very large number of degrees of freedom. This can be reduced by assumptions regarding classes of equals considering the underlying specification and the behaviour of the associated elements. For instance, an engineer could assume that the shortest duration of a value remaining in a channel is zero. Or he could say that the equality solver is allowed to assume zero as the minimum time of remaining in a state, if there is no other constraint resulting from the equality system

Considering our example again, the trial to solve the equality system leads to the following result:

- equals (15) and (16): $t_{3.1}(k) = k*100 - 10$ (29)
- equals (14), (27): $t_{3.2}(k) = k*100 +/- 10$ (30)
- equals (18), (29), (30): $s_{3.2} = \text{interval}(0, 20)$ (31)
- equals (17), (29), (30): $s_{3.1} = \text{interval}(100, 80)$ (32)
- equals (15) and (24): $t_{5.1}(k) = (k+1)*100 - 10$ (33)
- equals (14), (28): $t_{5.2}(k) = (k+1)*100 +/- 10$ (34)
- equals (25), (33), (34): $s_{5.2} = \text{interval}(0, 20)$ (35)
- equals (23), (33), (34): $s_{5.1} = \text{interval}(100, 80)$ (36)
- equal (15): $t_6 = k * 100 - 10$ (15)
- equals (15), (26): $s_6 = 100$ (37)

The remaining part of the equality system consisting of the equals (19), (20), (21) and (22) is not solvable any more, because of having more degrees of freedom than equals. In order to reduce the large degrees of freedom, the following, above mentioned assumptions are made:

- It is assumed, that for inspecting the worst case the duration of values remaining in the channel is zero. This leads to intervals as:
  - intern_inp = interval (not-known , 0) (37)
  - intern_out = interval (not-known , 0) (38)
- It is assumed, that the shortest run time to be allowed for a procedure is zero. This assumption makes the succeeding verification steps easier. For instance, it's easier to assure, that in a system with several parallel tasks, a procedure will be ready within the interval [0 , 100], than at the earliest time 50 ms after starting and at least after 100 ms. This leads to the assumption:

  $s_{4.2} \quad = \quad \text{interval} (0 , \text{not-known})$ (39)

Thus, the remaining part of the equality system is solvable by:

- using the worst case of the equal (37) (intern_inp=0) combined with the worst case of (19) leads to:

  $t_{4.1}(k) \quad = \quad k*100 + 10$ (40)

- using the worst case of equal (38) (intern_out=0) put into (22), it leads to:

  $t_{5.1}(k) \quad = \quad (k+1)*100 - 10$ (41)

- equals (40), (41), (39) and (22) lead to:

  $s_{4.2} \quad = \quad \text{interval} (0 , 80)$ (42)

5.4 The Result

Figure 8 visualizes the temporal behaviour of the software design with respect to the temporal requirements of the conceptual design. The strategy presented above leads to intervals with upper limits, based on the upper limits of all other intervals of the inspected sequel.

The simplicity of the result may surprise, compared with the effort being necessary to reach it. But though the only roughly presented solution concept is a formalized and generalized one, the automation of the whole solution process is possible. Applied to a less trivial example, we're sure of having a very powerful and helpful tool actively supporting the development process and leading to correct temporal requirements in the software design.

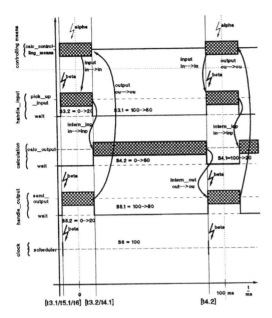

Figure 8: Temporal behaviour of the complete real-time system on the level of the software design.

6. Conclusion

What did we achieve now with our approach from a practitioners' point of view?

- With the knowledge about the internal time relations in the software design, the engineer is now able to make his further design decisions - for example defining the priorities of the tasks, or deciding what scheduling algorithm would fit best the (now exactly known) quantitative time requirements - on a very precise and correct information basis.

- There further exists a solid basis for the verification, that in spite of performing parallel tasks, the time requirements 'mapped' into the software design (out of the time constraints of the conceptual design) will be met by the estimated cpu-times of the specified elementary functionalities. The estimation of the cpu-times could possibly be based on special compilers as already mentioned in section 2.

- For testing the meeting of the time requirements at run time, we do know now exactly where to place the certain time conditions in the code. If all run time tests give positive results, the time requirements will be met, otherwise an exception management could be started.

- After the verification process we gained the following assertions referring to several formal specification parts:

  - The formal specification parts have no deadlock and no lifelock.

  - The specification parts are complete with respect to the control flow.

  - The constructed time requirements are correct in spite of the different structures of the software design and the conceptual design. To say it more generally, our approach allows us to assert the correctness in spite of the fracture in the structure between the two concerned specifications.

- Finally, with practitioners', formal methods generally have the reputation of being not useable for real applications. We think that we have shown with our approach that even after the formalization of the development

method based on the well known state processes, the understandability of the method hasn't decreased. Moreover, we can draw the conclusion, that the connection of formal methods (including the formal verification) with the already practically proved methods of the software engineering domain, leads to a really powerful method.

- The concept not only allows the assuring of the correctness between two specifications, but also the aiding of development of the correct specification, activly.

7. Apology

Readers of this paper whose native tongue is English, will certainly notice numerous grammatical and spelling errors, plus sometimes a rather strange writing style. As the author's native tongue is German, he apologizes for being unable to use a 'perfectionist approach' in writing this paper in English. He would appreciate an 'error-tolerant' attitude of the readers.

Literatur:

[Agne91] Agne, R.: An Approach for Guaranteeing temporal Behaviour in Distributed Real-Time Systems (in German: Ein Verfahren zur Garantie des zeitlichen Verhaltens in verteilten Echtzeitsystemen), Proceedings of Control Computer Systems '91 (In German: Tagung: Prozeßrechensysteme '91), Berlin, Informatik Fachberichte Nr. 269, Springer-Verlag, pp. 80-89

[Berz88] Luqi and Valdis Berzins: Rapidly Prototyping Real Time Systems. IEEE Software, September 1988, pp. 25-36

[BrMa86] Bruno, G. and Marchetto, G.: Process-Translatable Petri Nets for the Rapid Prototyping of Process-Control Systems IEEE Trans. on Software Engineering Vol. SE-12 (Febr. 1986) No. 2 pp. 246-357

[HLNP90] Harel, D., Lachover, H., Naamad, A., Pnuelli, A., Politi, M., Sherman, R., Shtull-Trauring, A., Trakhtenbrot, M.: STATEMATE: A Working Environment for the Development of Complex Reactive Systems, IEEE Trans. on Software Engineering, Vol. 16, No. 4, April 1990, pp. 404-413

[HaPi87] Hatley, D. J. and Pirbhai, I.: Strategies for Real-Time System Specification, New York 1987, Dorset House

[JaLe89] Jaffe, M. S. and Leveson, N. G.: Completeness, Robustness, and Safety in Real-Time Software Requirements Specification, Proceedings of the 11th International Conference on Software Engineering, 1989 Pittsburgh, Pennsylvania, May 15-18

[JLHM91] Jaffe, M. S., Leverson, N. G., Heimdahl, M. P. E., Melhart, B. E.: Software Requirement Analysis for Real-Time Process-Control Systems, IEEE Trans. on Software Engineering, Vol. 17, No. 3, March 1991, pp. 241-258

[Laub89] Lauber, R.: Forecasting Real-Time Behavior During Software Design using a CASE Environment, The Journal of Real-Time Systems 1, Kluwer Academic Publisher, Boston, 1989

[MeWa86] Mellor, S.J. and Ward P. T.: Structured Development for Real-Time Systems, Volumes 1 - 3, Prentice-Hall, 1986, ISBN 0-13-854803-X

[Ostr89] Ostroff, J. S.: Temporal Logic for Real-Time Systems, Research Studies Press LTD, England 1989, ISBN 0 86380 086 6

[OsWo90] Ostroff, J. S. and Wonham, W. M.: Framework for Real-Time Discrete Event Control, IEEE Transactions on Automatic Control, Volume 35, No. 4, April 1990

[Roes79] Roessler, R.: Operating systems strategies to cope with time problems in process automation (In German: Betriebssystem-strategien zur Bewältigung von Zeitproblemen in der Pro-zeßautomatisierung). Ph.D.Thesis - Dissertation, University of Stuttgart (1979)

[RSS90] Ramamritham, K., Stankovic, J.A. and Shiah, P-F.: Efficient Scheduling Algorithms for Real-Time Multiprocessor Systems, IEEE Transactions on Parallel and Distributed Systems, Vol. 1, No. 2, April 1990

[Stoy87/1] Stoyenko, A.D.: A Real-Time Language with a Schedulability Analyser, Ph.D.Thesis, Department of Computer Science, University of Toronto, 1987, available as Computer Systems Research Institute Technical Report CSRI-206

[Stoy87/2] Stoyenko, A.D.: "A Schedulability Analyzer for Real-Time Euclid", Proceedings of the IEEE 1987 Real-Time Systems Symposium, December 1987, pp. 218-225

A METHOD FOR COMPUTING HOL SW TIME RESPONSE AND ITS VALIDATION

G. Cantone*, E. Ciancamerla** and M. Minichino**

*Dept., of Information and Systems Engineering, University of Naples "Federico II", via Claudio 21, 80125 Naples, Italy
**Dept., of Energetics, ENEA CRE Casaccia, via Anguillarese 301, 00100 Roma A.D., Italy

Abstract. The paper describes a computational method to assess time response of critical life program units, presents some validation results and gives some implementation guidelines for a related tool. A complete environment currently used in industrial applications, has been used to compare software timeliness, computed by the proposed method, versus time measurement, typically performed along software development process.

Keywords. Computer software, computational method, microprocessors, models, real time computer systems, safety, software tools, time computation, time measurement.

INTRODUCTION

This paper[1] deals with a method to assess and validate temporal properties of Higher Order Language (HOL) SoftWare (SW) as part of Real Time (RT) systems, by a computational model. A basic requirement of a RT system versus other computer based systems is to properly implement its specific functions according to its timing constraints (Del Corso, 1989). From timing assessment point of view, the SW subsystem is perhaps the most critical one, both for its major timing contribute to the overall system timing budget, and for its eventual unforeseeable behaviour. Particularly it has to be guaranteed that each task duration is inside its specified time boundaries. Moreover the knowledge of the SW temporal properties should be abstracted, respect to hardware and compiler details, and anticipated as possible. An early knowledge allows to identify proper SW architectural choices in order to fit SW timing requirements; besides it allows prompt maintenance of products by detecting and suppressing time bottlenecks, enhancing products and adapting them to new environments.

Temporal predictability of tasks has been largely studied and some proposal of temporal verification of RT systems are known from novel scientific literature. Static prediction of sequential programs timing was discussed by (Mock, 1984), (Cantone, 1987),(Koza, 1989), (Shaw, 1989). Prediction methods are essentially based on source code decomposition into basic items. Each item is then transformed into a duration say by seeking its execution time from an apposite repository. Timing prediction of concurrent processes was discussed by (Mock, 1987), (Henzinger, 1990), (Shymasundar, 1990). Finite State Automata and temporal logic were used in order to respectively study the Maximum Parallel and the Interleaving Models.

A HOL method allows to assess any SW worst/best timeliness by computing and combining basic, elementary or complex, HOL block timeliness. The elementary durations can be evaluated or measured and then stored into an apposite repository, once for ever. The concept of repository can be extended from elementary blocks to complex ones. Infact, previously evaluated durations of complex blocks can be reused in order to compute time behaviours of more complex blocks up to applications. The ability of the HOL model to be abstracted with respect to the coding language and the hardware level allows to overcome some dependencies from language evolutions and hardware migration imposed by technology, just including time variations due to language changes into upper level repositories and time variations due to the hardware upgrading into lower level repositories. The above sketched HOL method covers both sequential and concurrent SW time assessment. This paper deals with an HOL method for time assessment of sequential SW and its validation versus conventional timing assessment methods.

TIME RESPONSE BY MEASUREMENT

Usually, the development of RT software, both critical and not critical, is practically performed by using environments which are able to support:

* development and verification of a product, namely the Host

[1]The work underlined by this paper was partially supported by the "Progetto Finalizzato Sistemi Informatici e Calcolo Parallelo" (PFSICP) of CNR under grant n.89000-52-89

Product, (HP), by the use of all the host available facilities;

* transfer of HP from the host environment to an emulator and set up and verification of the product in the new environment. A new product, namely the Emulator Product, (EP), is obtained. The emulator is typically characterized by having the same processor of the target system, some amount of memory, typically RAM, auxiliary input/output units for debugging and testing the product;

* transfer of the final product, namely the Target Product (TP), to the target EPROMs, and finally to the target system.

In the above synthetized process, time assessment and evaluation goes throught three phases. At the beginning, the product timing is the timing simulated by the Host environment. Timing is disclosed in the following way: the product, generally expressed by assembler or by a traditional HOL, is instrumented by some checkpoint instructions that can be interpreted by the simulator of the target processor. Generally the simulator, executes the machine code of the target processor and gives auxiliary information, as the simulated execution time. Corrispondently to the execution of checkpoint instructions, the simulator memorizes the value of the simulated time. Differences between simulated times represent simulated durations. The above scketched temporal verification method, though requires executions, is typically static; the implementation of the simulated time is done by tables which associate each source instruction to the expected target execution time.

Time assessment of EP can be performed with more thinness. A time analyzer system. with no intrusion on the EP, can be connected to the emulator processor and time can be measured very precisely. Such a system includes S/W utilities and an ergonomic user interface that allows to select program blocks or paths to be measured, to obtain results, in a single or cumulative way, both in grafic and numeric representations. The same system can be used for the TP final assessment.

The above synthetized approach to time assesment presents some limits, need a lot of human work and can result not fully applicable for verifying basic properties of advanced critical hard real time applications, like aerospace, nuclear or medical applications. Some drawbacks can be synthetized in the following points. The above approach and its tools basically refer to:

* applications run by a small number of processors, typically one: to increase reliability and/or performances, advanced critical applications need fault tolerance, multiprocessors and multicomputing;

* a single or a small number of diverse processor types; the ones for which exists a host simulator, emulators and time analyzers which are appropriated for the target processor. Advanced critical applications could need special or new processors. New hardware based anlyzers/emulators and host simulators could be expensive and, more important, not available when needed;

* poor use of structurization concepts, typically limited to instruction labels; that means manual work is needed to request and structure temporal results.

Another way to predict SW timeliness is based on temporal models, which tray to express time formally by closed expressions. These methods could expecially allow to cover the early phases of SW life cycle as requirement and specification phases. Tools, based on such models, are offerted to detect improper temporal behaviour of SW (deadlocks, wrong sequences of concurrent processes, time responses). However, at the present state of the art, the complexity introduced by these methods seems to be even higher then the complexity of the SW system under prediction.

THE TIME COMPUTATIONAL METHOD

Let's consider first HOL program units. Istances of such units are compilation units, as Packages, Functions and Procedures in the ADA jargon, un-compilable upper units, as Generic Packages, or lower units, as Declare Blocks and code fragments. Firmware and Assembly level units will be treated later-on. Particularly, let's begin by referring to units U, which internal control flow can be expressed by using a few structured constructs, say sequence (S), if-then [-else] (I) and while-do (W) constructs. A programming unit U, in general, exports one or more entry paths. Said U an entry of U and U_0 the U's upper nesting level, U_0 can be decomposed as following:

Case U_Type (U_0)

When S $\implies$ $U_0 := U_1 \to U_2 \to ... \to U_{last}$;

When I $\implies$ $U_0 := (ITP \to UT) \# (IFP[\to UF])$;

When W $\implies$ $U_0 := (WTP \to UL)^I \to WFP$;

End Case;

Where P stands for for Predicate, T and F for True and False, respectively, L for Loop counter size. Characters # and $\to$ denote the diramative and the total ordering operators, respectively. The notation $"^I"$ is mutued by the Kleene operator and square brackets bound optional parts. The described decomposition criteria can be recursively applied to nested units up to basic units.

Let TE be a trasformation of U into an expression of the duration of U:

TE(null) := 0.0; TE(K) := K, $\forall$K natural;

TE(a $\to$ b) := TE(a) + TE(b);

$TE(a^i) := I * TE(a);$

$TE(a \# b) := TE(a) \# TE(b).$

By applying TE to the above decomposition of U, it derives:

Case U_type (U_0)

When S $\implies TE(U_0) := \sum_{i=1}^{last} TE(U_i);$

When I $\implies TE(U_0) := (TE(ITP) + TE(UT)) \# (TE(ITF)$
$+ TE(UF))$
$:\simeq TE(IP) + (TE(UT) \# TE(UF));$

When W $\implies TE(U_0) := I * (TE(WTP) + TE(UL)) + TE(WFP)$
$:\simeq I * (TE(WP) + TE(UL)) + TE(WP);$

End Case;

Thus a temporal expression generator can be recursively defined whenever a Basic unit Type (BT) is used in order to complete the recursion on U. The generator returns the identifier of U, if U is a BT unit, or an expression of unit identifiers, in the general case. A repository is used to transform BT unit identifiers into their duration expressions. Structures of BT units, in turn, can be identified and saved into, and fetched from a properly structured data base repository.   Temporal expressions, as defined above, hide dependencies from HW/SW environment which translates and executes the related SW.

Let's look at, now, the static trasformation of a temporal expression into a proper finite temporal duration value. For this aim some constrains must be met and some details have to be specified. In particular:

* the upper Worst Response Time paradigm is assumed to observe SW temporal behaviour;

* loop counter bounds are requested to be statically defined;

* the basic environment or, better, any diverse basic environment, which is used by the RT application, is requested to present deterministic behaviours with respect to durations of basic units that it is charged to handle.

Under the above assumptions, let's denote by D a function which transforms a basic unit into its duration. The implementation of D depends on the chosen behavioral paradigm. For instance, if the Upper Bound (UB) RT paradigm is chosen, then the operator "#" can be suppressed, letting any temporal expression assume anywhere a traditional algebric form:

$UBRT(TE(S)) := TE(S);$

$UBRT(TE(I)) := TE(IP) + max(TE(UT), TE(UF));$

$UBRT(TE(W)) := TE(WP) + max(I) * (TE(WP) + TE(UL))$

Morover the implementation of D depends on some characteristics of the basic unit set. In particular, D can be concerned with static or dynamic set of basic units. To improve the adaptability of D to changes of the executing environment, D is expected to comply with a configurable set of basic units. To improve the comparability of D's results with respect to diverse hardware, language, compilers and basic units and to let D support dependable applications based on diversity, D is expected to comply with families of environments. Concerning structurization of an application, D is expected to comply from assembly to advanced HOL basic units. Finally, to improve the reusability of RT SW and the related experiences, temporal expressions and durations are expected to be saved together the related units and packaged with results of human experience (Basili, 1991).

In order to give some frameworks of possible implementations of D, let's assume that a layered repository is used to store basic units identifiers and durations. Independently from the level of the basic unit we refer, its repository entry could be updated by measurement. A hardware indipendent method for such an automatic measurement by the target environment was discussed by (Cantone, 1991). To conclude this section let's give some instances about the work that must be done whenever some changes occur in some object of the target environment, in order to adapt the repository of the basic units to the new conditions.

a: Instances of changes

a0.1: firmware of a processor;
a1.1: technology of an handshaking register;
a1.2: sensor or actuator;
a1.3: processor clock rate;
a2.1: basic hardware machine;
a3.1 compiler;
a3.2: compiler optimisation degree;
a4.1: instruction;
a4.2: fragment of code;
a4.3: implementation (of a unit);
a4.4: code language;
a5.1: module interface;
a6.1: enclosed new module.

B: Instances of repository updating

* BF: Firmware level Basic units; a0 changes affect directly the content of the Firmware Repository (FR). For other consequential changes see BA.

* BA: Assembly level Basic units; a0 and a1 changes affect directly the content of the Assembly Repository (AR); a2 changes affect the ASM based temporal expressions. For other consequential changes see BH.

23

* BH: <u>HOL Basic units</u>; changes from a0 to a3 address for updating the HOL repository and for revaluating the values of the temporal expressions. Moreover any a4 change also addresses for re-valuating the temporal expression of the affected unit. Furthermore, a5 changes require the re-evaluation of temporal expressions of units which are connected to the affected unit. Finally, a6 changes require an expansion of the HOL repository by adding the new module to the apposite HOL layer.

VALIDATION ENVIRONMENT

In order to validate the described HOL computational method, a complete environment, currently used for industrial and safety related applications, has been used to compare SW timeliness computed by the above method, versus time measurements typically performed along SW development process. The environment, developed by ENEA/Westinghuose and others (Bologna, 1988) and modified by ENEA for this specific purpose, allows time measurement according to traditional methods, covering all SW development phases, as described in previous sections. It includes time measurement systems both on host computer and on target system. Particularly it was tailored to Intel tecnology.

As host machine, a computer of VAX DIGITAL family has been used and the time measurement system, under VMS operating system, includes automatic tools to support the development and the simulated execution of the software under computation. Besides, the system allows to automatically report the results, and to have a complete traceability of the measurement phases, that is expecially required from current guidelines concerning SW of safety related applications. The environment refers to Intel PL/M language which allows to obtain acceptable time response joined to structurization sintax. Sample codes, were choosed following the requirements indicated next section. Automatic tools that support SW development, generate, in a semi-automatic way, a "pilot" code that will invoice the generic block to be measured in terms of temporal behaviour, and produce the executable code. Then the number of clocks elapsed between blocks of instructions is computed and highligthed, by a simulator; the duration of the block is computed through the difference of the clock numbers between the block start point and the block end point, marked by labels. All the automatic tools are integrated and activated by an unique user interface that allows to perform time measurements, step by step, in a controlled and traceable way.

The first part of Fig. 1 shows the host environment in detail. The software system is represented by a set of PL/M modules and their relashionships. A module is a collection of procedures, elementary or composed by other procedures. The host measurement system

asks for a module as input and performs the following steps:

1. Extracts, from the module, the single procedures on which the time measurement has to be performed (Parser).

2. Builds a pilot code that will call the procedure for the needed set of measurement cases (Builder).

3. Asks for the set of measurement cases, guiding and helping in inserting data (Test generator).

4. Generates a special code to manage the measurement cases in order to produce automatically, during the execution of the program, the report of results and the trace of execution (Test generator).

5. Compiles, links and locates the program, keeping trace of all operations.

6. Executes the program by the simulator.

7. Produces result files, particularly files containing the values of clock that corrispond to interesting points in the program and files containing information on program correct/uncorrect execution.

Time measurement enviroment on target computer allows to evaluate time reponse of SW subsystem up to the system prototype. Host and target measurement environment are linked by an asyncronous data link. The target enviroment includes basically two subsystems, as shown in the second part of Fig. 1:

i) an In Circuit Emulator (ICE) for xxx86 Intel microprocessors hosted by an Intel development system and

ii) a timing analyzer, with a probe for xxx86, installed on 386 Personal Computer (PC).

The two subsystems are linked together by a Xon/Xoff protocol.

The SW code under measurement is downloaded from VAX to the development system. The symbol table of the code is passed to the timing analyzer. The code can be executed on the development system, on target prototype by using ICE, or, at the final step of the development process, on target prototype through target EPROMs. The analyzer, which is equipped by an appropriate probe to be plugged on the target prototype microprocessor socket, allows to measure time durations between labels, selected from the code symbol table. The use of the described target environment is restricted to prototypes based on the same ICE and SW analyzer microprocessor probes. The use of SW analyzer is not restricted by SW language.

However the sample codes, used to validate the method and shown next section do not require I/O devices and links with the external world, but they require just to be hosted on an xxx86 Intel microprocessor. So the complete measurement target environment, required for more complex RT software, is not needed. A xxx86 Personal Computer was choosen as target system and its facilities were used in order to compute time durations of sample codes on the target system.

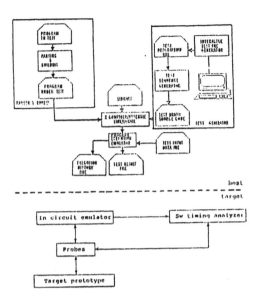

Fig. 1. Host-target time measurement environment

VALIDATION RESULTS

The proposed method is based on temporal expressions associate to HOL typical structures. The validation phase was performed experimentally verifying the following ordered steps:

1. algebraic assumptions/properties;

2. basic structures

3. sample units

From the previous sections we assume:

* a method and related tools, ($\underline{M}$), to measure the execution time by using an assigned environment (E) of a code unit (U) under any data (d) belonging to the input domain (ID_u) of U:

$$\underline{M}: \forall U, \forall d \in ID_u \xrightarrow{E} \underline{M}(U(d))$$

where $\underline{M}$ is expressed by some time unit, say the number of E clock ticks;

* a deterministic environment:

$$E: \forall U, \forall d \in ID_u, \partial \underline{M}/\partial t = 0.0;$$

* an environment indipendent duration expression of U (TE):

$$TE: \forall U, \ TE(U) = TE(U_1, U_2, ..., U_{last})$$

where any (U_i) corrisponds to a basic or complex unit of U and TE is a composition of operands, like "#", "+", "()^1";

* a method D to compute TE in E.

Our aim is to verify that the "Property of Consistence":

$$-(U,d), \ \underline{M}(d) = D(TE(\underline{M}(U_1(d_1)), \ \underline{M}(U_2(d_2)), ..., \underline{M})(U_{last}(d_{[last]})));$$

holds true.

Some experiments were designed and performed to verify such a Property both by the host and by target environment. Particolarly Units (U_s) were selected, from sample codes typically used for comparing compiler performances, in order to verify basic properties, recursive extention included, of temporal operators ("#", "+", "()^1" we assumed above.

A basic experiment was performed to verify the additive property of durations, by a unit computing the distance between two points (x_1, y_1), (x_2, y_2).

Basic Experiment n.1

test case 1:	U computes distance between 2 points;
	$U_1: Q := (x_2 - x_1)^2 + (y_2 - y_1)^2;$
	$U_2: \mathrm{sqrt}(Q);$
expression :	$TE(U) := TE(U_1) + TE(U_2)$
d :	($x_1 := 4.3, y_1 := 7.8$); ($x_2 := 1.0, y_2 := 12.0$)

	U(d)	$U_1(d_1)$	$U_2(d_2)$
$\underline{M}$:	4101	1369	2740
computation:			4109

In the assumption of the additive property, by comparing computed and measured durations, an error results. In order to understand the reason of such an error and complete basic experiment n.1 the following test cases were performed.

Basic Experiment n.1 (cont')

test case 2:	UD = sequence(test case 1 & test case 1)
expression :	$TE(UD) := TE(U) + TE(U)$
d:	as test case 1
$\underline{M}$:	5472

test case 3:	UT = sequence(test case 2 & test case 1)
expression :	$TE(UT) := TE(UD) + TE(U)$
d:	as test case 1.1
$\underline{M}$:	8204

Measures from experiment n.1 imply:

$$\mathrm{abs}((\underline{M}_i - \underline{M}_j)/(i - j)) = \underline{M}_1(U) - \mathrm{Err}, i \neq j;$$

from which Err=8 is a systematic error, may be due to the clock simulator interference.

Similar experiments were carried out in order to validate the temporal model consistency with repect to chained and nested "#" and "()$^{\bullet}$" operators. Moreover, experiments were extended to include more powerful programming constructs, as "For", "Case", "Multi-Exit" Loops. Following, results from a cumulative experiment are reported and discussed. They refer to the "Bubble Sort" test case and report error filtered values. Figure 2 shows Bubble Sort program sample; comments are used to identify units.

25

```
FixSort: DO;
DECLARE elencoing(10)-
-INTEGER INITIAL (10,9,8,7,6,5,4,3,2,1);
DECLARE (Jj,save,n) INTEGER;
DECLARE (count,riemp) INTEGER INITIAL(1);
DECLARE true LITERALLY '0FFH',false '0';
DECLARE (scambio,cond1,cond2) -
-BYTE INITIAL(true,true,true);
Sort: PROCEDURE;
DO WHILE (scambio)AND(cond1); /* unit WP1, maxloops:=n-1,
                   loop_counter is count */
    scambio=false;        /* unit U₂ */
    n=riemp-count;
    DO J=1 TO n;         /* unit FP3, maxloop=n-count */
        cond2=(elencoing(J-1) > elencoord(j));/* unit U₄ */
        IF cond2 THEN;       /* unit IP5 */
            DO;        /* unit U₆ */
            save=elencoing(J-1)
            elencoing(J-1)=elencoord(j);
            elencoord(j)=save;
            scambio=true;
            END;
    END;
    count=count+1;       /* unit U₇ */
    cond1=(count<riemp);
END;
```

Fig. 2. Bubble sort program sample

Let's begin by taking in consideration results we obtained by the host environment.

Cumulative Experiment:

test-case U: Bubble Sort *(see Fig. 2);
Sub-units: $U1=WP1,U_2,U_3=FP3,U_4,U_5=IP5,U_6,U_7$;

*(see comments in Fig. 2);

worst path expression: $TE_w(U) \sim TE(WP1)+TE(U_2)+(n-l)*$
$(TE(FP3)+TE(U_4)+TE(IP5)+TE(U_6)+TE(U_7))+TE(WP1)$;
$TE_w(U)= TE(WFP1)+(n-1)*(TE(WTP1)+TE(U_2))+(n-1)*$
$TE(InizFP3)+(n-2)*TE(UT7)+TE(UF7)+n*(n-1)*(TE(FP3)+$
$TE*(U_4)+TE(IPT5)+TE(U_6)+TE(U_7))/2+(n-1)*$
$(TE(IFP5)-TE(ITP5))$;
d_w:n=10; inversely sorted array of integers;

Results:

Item	Value	Source
U	19120	measurement
WPT1+U2	114	measurement
WF1	42	repository
Iniz_PF3	56	repository
PF3	75	repository
U4	94	measurement
ITP5	24	repository
IFP5	27	repository
U6	179	measurement
UT7	87	measurement
UF7	85	measurement

Computation

$D_w(U)=9*114+42+9*56+8*87+85+45*(75+94+24+179)+9*(27-24) = 19120$;

$M(U(d_w)) - D((U(d_w)) = 0$

The consistency of the proposed method was also verified in the target environment, by using some measurement tools. Next figures show results obtained by using a tool (Cantone, 1991), based on the "software amplification" of the duration to be measured. Figure 3 shows differences between measured and computed durations of a

while loop versus the duration amplification rate. From Fig. 3 results that such a difference can be easily constrained to be less then 0.05 microsec. So the proposed method can be directly used also to init the time repository of basic HOL units, particularly when the basic hardware machine is not modelled and simulated or when the code can be accessed for compilation and execution, not for reading.

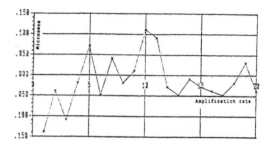

Fig. 3. While Loop (Differences between Measured and Computed Durations)

Figure 4 shows results we obtained by using information from such a repository in order to compute worst duration of a bubble_sort procedure when run by an Intel target environment.

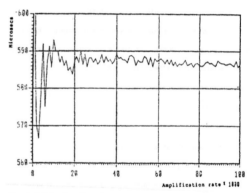

Fig. 4. Bubble Sort (Duration by Computation)

CONCLUSIONS

The experimental work showed that the HOL method can be used to assess software timeliness, overcoming some dependencies of typical time measurement environment from language evolution and hardware migration imposed by technology.

The effort described in the present paper just represents a step, necessary for a more detailed definition and validation of an extended HOL method, for timing assessment of concurrent SW, closer to RT systems.

The ultimate long term objective is to produce a design support enviroment which is well-populated with repositories of basic SW elements and of ready-made SW components, and automatic tools

In order to compute automatically the RT SW timeliness by using the information contained in the repositories. The automation level of such an environment is restricted, and, at the present time, human experience is requested to establish effective SW worst paths for computing actual best/worst cases.

REFERENCES

Basili V.R., G. Caldiera, G. Cantone (1991). A reference architecture for the component factory. UMDCS and UMIACS Technical reports, C.P.(MD).

Bologna S., G. Carra, E. Ciancameria, and S. Ratto (1988). An industrial test environment for safety critical software testing. Proc. of AICA annual conference, Cagliari (I).

Cantone G., and A. Esposito (1987). An initial approach to response time verification of critical programs. Proc. of CSCI-CNR int. conf. on massively parallel systems, Naples (I).

Cantone G., and F. Gragnani (1991). Methods for the assessment of the temporal behaviour of program and code fragments. SP6-PFSICP-CNR Report.

Del Corso D., F. Maddaleno, M. Minichino, and E. Pasero (1989). Resinchronization interfaces: sources of metastability errors in computing systems" Proc. of International Conference on Dependable Computing for Critical Applications, CA (USA).

Koza C., and P. Puchner (1989). Calculating the maximum execution time of real time programs. The journal of Real Time Systems, vol. 1.

Henzinger T.A., Z. Manna, and A. Pnuely (1990). An interleaving model for Real Time. Proc of JCIT, Gerusalem (IS).

Mock A. K. (1984). Design of RT programming systems based on process models. Proc. of RT Systems Symposium, IEEE press.

Mock A. K. (1987). Systesis of real time message processing systems with data driven timing constrains. Proc. of IEEE-RTS.

Shaw A.C. (1989). Reasoning about time in the higher level language software. IEEE-TSE, vol. 15, n. 7.

Shymasundar R. K., and L. Yuhsiang (1990). Static Analysis of real time distributed systems. IEEE-TSE, V. S16, n. 4.

IMPROVING SOFTWARE QUALITY IN A SPACE APPLICATION

A. Pasquini

ENEA, Via Vitaliano Brancati 48, 00144 Rome, Italy

Abstract. Computerized systems are getting more and more used in spacecraft, both for mission control and measurement equipment control. But space missions require high investments and sometimes cannot be delayed or repeated because are related to special natural events. For these reasons a computerized system failure could produce irreparable consequences as in the Phobos I case. In this paper an activity of software quality improvement and software reliability evaluation is presented. The activity is presently in progress and concerns software that will be used to control a measurement equipment during a Russian-European space mission on Mars. A brief description of the mission and of the measurement equipment is given. Then the adopted methodologies of fault avoidance, and detection and of failure detection and containment are analyzed together with their selection criteria. Finally the paper presents the software reliability evaluation activity that will be performed using an experimental model especially developed for critical application.

Keywords. Aerospace computer control; program testing; reliability; software development; software engineering; software reliability; software tools.

I - INTRODUCTION

Computer systems are beginning to affect nearly every aspects of space missions. The applications range from the control of the spacecraft to that of the communications and of the measurement equipments used for scientific experiences (Ceruzzi, 1988). All these applications require that the computer systems are able to perform their functions in the specified use environment. Indeed, a failure of these systems could have serious consequences since space mission requires high investment and sometimes cannot be delayed or repeated because are related to special natural events. Furthermore, there is no possibility of a direct operator control and also the chances of intervention from earth are often limited (Ceruzzi, 1988; SCIENCE, 1989).

Unfortunately this field of application presents several additional difficulties that affect the design and development techniques of computer control systems. Power consumption, space occupancy and weight of the control system must be frequently kept under very restrictive limits and this affects the chance of adopting redundant configurations and diversity. Flux of electrons, protons and heavier ions (cosmic rays) can produce effects called single event upset (SEU) and latchup. As a consequence RAMs content can be erased or corrupted (Benson, 1990; Pélegrin, 1988) and software means, like checksum, are not always sufficient to avoid the consequences of these failures (Spencer, 1990). Spacecraft development schedule and launch dates do not allow delays during control systems design and the haste imposed by looming deadlines could cause severe consequences (Lehenbauer, 1990). Spacecraft design is an on-going, iterative process and changes in the control system requirements are very frequent during the early stages of the development. Finally some of these systems, especially the control systems of measurement equipments, are usually developed by several teams from different institutions. In some cases these teams are also from different countries and sometimes with little knowledge of the software engineering techniques. The presence of all these interfaces would require a coherent management philosophy and methodology to guide the project to successful completion. But this is not always possible due to several reasons ranging from the lack of direct leadership to the physical distances between the teams.

The listed difficulties severely affects the quality of the control systems for this kind of application and several failures (Neumann, 1991) are present in the history of their development and use.

This paper presents an activity of quality improvement and reliability evaluation concerning software that will be used to control a measurement equipment during a Russian-European space mission on Mars.

Sections II contains a brief description of the space mission, of the scientific measurement equipment and of its development process. Section III describes the adopted methodologies of fault avoidance and detection and of failure detection and containment. It also describes their selection criteria. Section IV presents the software reliability evaluation activity that will be performed using an experimental model especially developed for critical application. Finally section V outlines the first conclusions and the lessons learned from this experience.

II - MISSION AND INSTRUMENT DESCRIPTION

Mars '94 is a soviet mission to Mars which aims to put two spacecraft in orbit around Mars. The launch date is foreseen in November 1994 and the arrival at Mars on September 1995.

Like in many other spatial missions, there is a strong cooperation of the soviet team with foreign scientific institutions. These institutions cooperate in the realization of the scientific instrumentation and will share the scientific results of the mission.

The instrument we are dealing with is a Michelson interferometer called Planetary Fourier Spectrometer (PFS). It is one of the so called "High priority" instruments of the mission together with an High Resolution Camera and an Infrared Spectrometer. These instruments will provide extensive information on the geophysical and geological processes which have modeled the surface of Mars. PFS will be developed in collaboration between two teams constituted by scientific institutions from USSR, Poland and ex DDR and Italy, France and Spain respectively.

The control system of PFS is constituted by several microprocessors working in parallel with different tasks. Its block diagram is shown in fig. 1.

There are four subsystems: Digital Arbiter Module (DAM) that has the general control of the experiment, it exchangedatabetween OBDM, ICM, Mass Memory and Telemetry and executes or sends the telecommands received from the earth control; Optical Bank Digital Module (OBDM) whose main functions are the data acquisition and transfer during the measurement session of the experiment, the control of the optical bank temperature, of the interferometer mirror movement, and of the sensors amplifier gain; Interferogram Compression module (ICM) which compresses data through Fast Fourier Transform; Scanner control which control the movement of the optical pointing system.

The four subsystems are not redundant because of the limited resources available on the spacecraft, especially power and space; for this reason all the reconfiguration capabilities of the system are delegated to software techniques as we will see in the next section.

III - TECHNIQUES ADOPTED

The main activities affecting the software quality were the improvement of the fault avoidance and detection and failure detection and containment techniques. A system operability analysis and a preliminary evaluation of the software and hardware reliability were used as input for the failure containment techniques selection and design.

Systematic analysis and design methods is one of the most effective techniques in avoiding faults since the early phases of the development. A systematic approach to the definition of the requirements and a more simple structuring of the data and of the software components are the most important advantages of this technique.

Several methodologies are available and the one adpted was selected on the basis of the project characteristics. In the following we describe and discuss the most important factors that influenced this selection.

- Scientific researchers, involved in the experiment, were chosen to design and implement the system. This people had not a software engineering background. This, and the strict deadlines imposed by the project schedule, limited the choice to the most simple and intuitive methods available.

- The organization of the whole space mission design and of the PFS design involved frequent modifications of the requirements. This problem was accentuated by the presence of several working teams from different institutions and different countries (see Section II): the project experienced some communication and coordination difficulties. The reasons mentioned above and the need of systems prototypes required the use of preliminary and sometimes not well defined requirements. Therefore, too expansive or time consuming development methodologies, such as formal development methods, were excluded. Further, changing requirements required a design technique able to animate the proposed system design, at least on paper, to verify their completeness and consistence.

- The system to be developed is a typical real time system with concurrent processes.

Considering all these requirements the structured development method proposed by Ward and Mellor,called Real-Time Yourdon (Ward, 1985), was chosen for the specification and design ofthe PFS control system. Several tools are available for the method and for its individual modeling techniques (NBS, 1982), but none of them was adopted. This decision was imposed by the strict deadlines: it would not have been possible to respect such deadlines considering the training for the tools use and the time needed to obtain the required hardware and software. However, this decision caused some serious consequences that will be described in Section V.

Fault detection techniques were integrated in a Verification plan, whose main components are: walkthrough, design animation and performance modeling for the design; functional testing and structural testing with path coverage analysis for the code (Myers, 1979).

Walkthroguh was selected because of its effectiveness in comparison with the little training required (Freedman, 1982; Weinberg, 1984). The lack of knowledge of the verification team was partially overcame using this technique and an experienced chairperson. Nevertheless, changing requirements obliged to several applications of the technique with the related waste of time and money and reduced its effectiveness.

Design animation was based on the features of the Real Time Yourdon development method. Its application produces several models of the system that can be used to simulate its behavior (Zave, 1984). Using this simulation it has been possible to verify the system analysis and design and to

prepare a Performance Modeling activity. The latter was required by the presence within the system of concurrent processes and of some functions with strict timing requirements like data acquisition or data formatting. Furthermore the concurrent processes have been sometimes implemented on different hardware, with the related communication and synchronization problems.

Failure detection and containment capabilities are based on the usual program diagnostics like variable range checks, configuration checks etc. (EWICS, 1990), on periodic checksum calculation of the program and on three level of time out checks. Software time out checks (first level) are used in program communications between processors where the sequence of actions is non-deterministic and during Direct Memory Access (DMA) data transfer when processors are in a hold state. Hardware time out checks (watch dog) are used to control both the critical task (second level) and the execution time of the whole working cycle (third level). In case of time out of the first level a functional recovery through re-starting of the interested tasks is attempted while in case of time out of the second or third level a complete re-start procedure is performed. Hardware time out and checksum are also required because the programs run from RAM,to compute to the highest possible speed, and RAMare very sensitive to radiation effects.

Failure containment is also based on the use of graceful degradation techniques (Sheridan, 1978) implemented at the subsystem level. As described in section II the system is constituted by several non redundant subsystems, performing different functions. Some of these functions can be completely or partially withdrawn without causing a complete failure of the system. For this reason DAM was designed in such a way that it can skip the data compression phase and directly guide the optical pointing system (with reduced performances). Then, DAM can afford a total failure of ICM or a partial failure of the Scanner. The functions duplicated within DAM and the functions that can be withdrawn were chosen on the basis of a system operability analysis and of an evaluation of the subsystems hardware reliability.

IV - SOFTWARE RELIABILITY EVALUATION

Several models have been proposed to estimate the reliability of software. References (Musa, 1987; Shooman, 1984; Yamada, 1985) contain a detailed survey of most of these models. In real applications only reliability growth models are widely used. These models estimate the number of errors remaining in a program and assume that their correction increases the reliability.

Unfortunately these models are not useful in this kind of application: their parameters are obtained from the testing history of the software using statistical consideration, therefore the confidence in the estimation grows with the size of the program and the number of faults detected. But, in critical applications, programs are usually of medium or small size and only a small number of faults is introduced during the development process. Therefore the confidence that they can provide is too low. Further, the realism of some of the underlying assumptions of these models is still questionable (Goel, 1985; Ramamoorthy, 1980).

The model that we are planning to apply to measure the reliability of PFS software is an input domain based model specifically developed for critical applications. It is presented in detail in (De Agostino, 1990) and assumes that:

$$R = 1 - \Sigma_{h=1}^{m} \int_{X_h} \text{Max}\{P(F_1)\mu_{XF_1}(x), \ldots, P(F_r)\mu_{XF_r}(x)\} p(x)\, dx$$

where: the program input domain X is divided in equivalence classes X_h (h = 1, 2, ... m); P(F) is the probability that faults are present in the program and $P(F_K)$ (K = 1, 2, ... r) the probability that faults of the k-type are present in the program; XF is the domain of the fault set F, it is divided into r-subset XF_k, each of which corresponds to a fault type F_k, and

$$U_{k=1}^{r} XF_k = XF;$$

$\mu_{XF_k}(x)$ is the probability that x belongs to XF_k; p(x) is the probability density function of the program input. In other words the model allows to combine the distributions of certain types of faults in the program and to weight them with the probability of the presence of each type of fault. In (De Agostino, 1990) suggestions and examples are given to guide the model parameter estimation.

The model is especially fitted for case in which a high level of reliability and high confidence in the estimate are required. It also provides indications of the number of tests needed to assure a predetermined level of reliability and of the optimal testing strategies for the program under evaluation.

Unfortunately it presents two drawbacks as well: the fault type distributions are obtained with strong assumptions on the fault characteristics and a significant effort is required for its application. For these reasons the model application will be limited to the most critical part of the PFS software and regarded as experimental.

V - CONCLUSIONS

The activity described in this paper is still in progress, therefore we can only outline the main difficulties and the lessons learned from this experience.

Section III describes the reasons why analysis and design tools were not adopted. Unfortunately this decision caused several consequences. The most common was the introduction of clerical errors in the design. Anyway most of them were found during walkthroughs or later development phases. But other consequences arose from the lack of the prescriptive method of work imposed by the tools combined with the "scientific" nature of the developers: it was difficult to apply configuration management and therefore to ensure consistency of the design deliverables during the frequent modifications required; furthermore, it was difficult to adopt and follow software engineering standards and therefore to ensure a more consistent approach to the development. The mentioned presence of several groups in the projects and the frequent up-date of the requirements increased these difficulties.

Even the effectiveness of walkthroughs was reduced by the frequent changes in the requirements since it was difficult to apply it to "frozen" development products.

For several reasons only the technical quality aspects have been afforded with the described activity. But, from the previous comments, it is possible to conclude that an equivalent effort in improving the project management is required to assure a full success in developing such systems.

VI - REFERENCES

Benson, D. B. (1990). Magellan spacecraft will need frequent guidance from Earth. ACM Software Engineering Notes, vol. 15, no. 2, pp. 22 - 23.

Ceruzzi, P. (1988). Beyond the Limits -- Flight Enters the Computer Age. Smithsonian National Air and Space Museum. Whasington, USA.

De Agostino, E., G. Di Marco, and A. Pasquini (1990). A Fault Domain Based Measure of Software Reliability. ENEA Internal report: ENEA RT 78/90, Roma, ItalEuropean Workshop on Industrial Computer Systems (EWICS) T. C. 7 (1990). Dependability of Critical Computer Systems III. P. Bishop (Ed.). Elsevier Applied Science, London.

Freedman, D. P., and G. M. Weinberg (1982). Handbook of Walkthroughs, Inspections and Technical Reviews: Evaluating Programs, Projects and Products. Third ed. Little, Brown and Co., Boston, USA.

Goel, A. L. (1985). Software Reliability Models: Assumptions, Limitations, and Applicability. IEEE Transactions on Software Engineering, Vol. SE-11, NO. 12, Dec. 1985.

Lehenbauer, K (1990). Software bug causes Shuttle countdown hold at T-31 seconds. ACM Software Engineering Notes, vol. 15, no. 3, pp. 18 - 19.

Musa J. D., A. Iannino, and K. Okumoto (1987). Software Reliability - Measurement, Prediction, Application. Mc Graw-Hill Book Company.

Myers, G. (1979). The Art of Software Testing. Wiley and Sons, NY, USA.

NBS (1982). Software validation, verification, and testing techniques and tool reference guide. NBS Special publication 500- 93. US Department of Commerce.

Neumann, P. G. (1991). Illustrative Risks to the Public in the Use of Computers Systems and Related Technology. ACM Software Engineering Notes, vol. 16, no. 1, pp. 2 - 9.

Pélegrin, J. M. (1988). Computers in Planes and Satellites. Proceedings of the IFAC Symposium SAFECOMP '88, Pergamon Press, Fulda, FRG.

Ramamoorthy C. V., and F. B. Bastani (1980). Modelling of the Software Reliability Growth Process. Proc. of the COMPSAC '80, Chicago, IL.

SCIENCE (1989). Phobos 1 & 2 computer failures. SCIENCE, vol. 245, Sept. 1989, p. 1045.

Sheridan, C. T. (1978). Space Shuttle Software. Datamation, vol. 24, July 1978.

Shooman, M. L. (1984). Software Reliability: A Historical Perspective. IEEE Transactions on Reliability, R-33 (1).

Spencer, H. (1990). Shuttle roll incident on January '90 mission. ACM Software Engineering Notes, vol. 15, no. 3, p. 18.

Ward, P. T., and S. J. Mellor (1985). Structured Development for Real-Time Systems. Yourdon Press.

Weinberg, G. M. (1984). Reviews, Walkthroughs, and Inspections. IEEE Transactions on Software Engineering, vol. SE-10, no. 1, January 1984, pp. 68 - 73.

Yamada, S., and S. Osaki (1985). Software Reliability Growth Modeling: Models and Applications. IEEE Transactions on Software Engineering, Vol. SE-11, NO. 12, Dec. 1985.

Zave, P. (1984). The operational versus the conventional approach to software development. Communications of the ACM, 27, no. 2, February 1984.

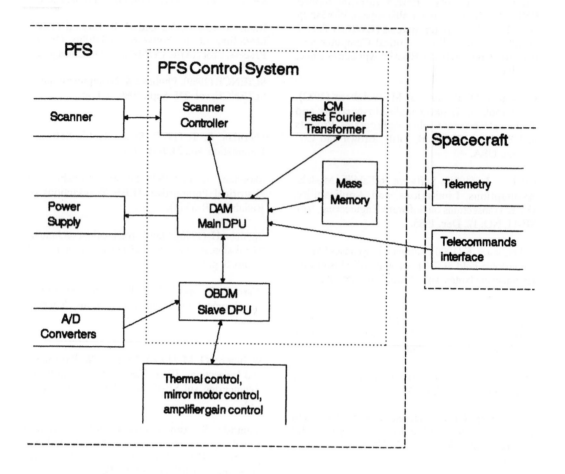

Fig. 1. PFS Control System architecure.

DARTS - AN EXPERIMENT INTO COST OF AND DIVERSITY IN SAFETY CRITICAL COMPUTER SYSTEMS

I. C. Smith, D. N. Wall and J. A. Baldwin

Control and Dynamics Dept., AEA Reactor Services,
AEA Technology, Winfrith, UK

Abstract. The DARTS project is an experimental investigation into the means of cost effective procurements of and the utilisation of diversity within safety critical computer based systems. The work concentrates on software issues where it is considered the major difficulties and uncertainties lie. This paper reports the way in which the project has been set up and in particular how the constraints and conditions of the experiment were arrived at.

Keywords. Computer Software, Computer Testing, Safety Software Development, Software Tools.

1. INTRODUCTION

Computer based systems are increasingly being used in safety critical applications where a very high degree of dependence is placed on the system such that their failure could lead directly to a loss of life. Some of these uses are overt, for example in the introduction of computer based systems for aircraft control and nuclear power plant protection, where they attract considerable public attention. Other applications of these systems are not so apparent, for example in: engine management and in medical equipment. In common with many computer based products there are many complaints about the poor performance and high cost of such equipment.

There are many problems associated with the development and use of these systems in the main these arise from the very high reliability targets to be demonstrably achieved. This demonstration activity is reflected in the high cost of introducing the systems, a key parameter in the decision making process as to whether to employ a computer based system.

The limited number and rather unique nature of safety critical systems makes it difficult to compare the effectiveness of different approaches to procurement especially in terms of exhibiting the features of interest, ie 'low' procurement cost and effective use of software diversity to enable claims of high reliability to be substantiated. One consequence of this is that much of the discussion of these system features requires significant extrapolations of the available information to be made to arrive at any conclusion. The DARTS project and experiment are intended to obtain improved data that can form the basis of a discussion of cost effective procurement of safety critical systems.

The paper starts by outlining the scope of the project and the approach adopted to obtaining the data on which to base the evaluation. The paper then described how the choice of target safety system, a computer based nuclear reactor protection system, was arrived at before moving on to describe the development methodologies to be investigated and the licensing and assessment framework within which the development takes place. A possible framework for acceptance

testing that might form part of system development process for ascertaining system release by supplier, system acceptance by customer and for assessment and licensing is outlined. The final sections of the paper describe the measures taken to collect process data and give the current status of the project.

2. THE SCOPE OF THE DARTS PROJECT

The prime objective of the DARTS project is to determine the most cost effective approach to moving from a requirements specification for a computer based system for a safety critical application to having that system installed and working. The aspect of system procurement of most interest is the software, as it is for this part of the system that the costs are subject to greatest uncertainty. Thus the prime objective reduces to establishing the methods and means of cost effective production, assessing and licensing of software. The concentration on software should not be seen as resulting in DARTS being a software engineering exercise alone as there are many issues and trade-offs to be considered. One example of this is the trade-off of cost of production against cost of assessment and licensing. These choices can introduce large uncertainties into system costs and timescales and must therefore be addressed in this exercise. Diversity might appear irrelevant in the above discussion, but it is becoming increasingly a feature of safety systems and has major cost implications. For example rather than procure a single system to meet a very high reliability target it may be cheaper to procure two less reliable, cheaper systems to meet the target. Here the treatment of diversity could be the key to the most economic approach.

The output of the DARTS project should be guidance on the means of obtaining a computer based system for a safety critical application in a cost effective manner. This output must include discussion and guidance documents on the production, assessment and licensing of such systems. The first topic should not just be yet another general document on good software engineering practice, but should address system issues and try to compare the consequences of adopting different approaches and tools. The assessment document likewise cannot be simply a guide to metrics that are available to measure software quality but again should give helpful direction on how to achieve a successful outcome. The last item, licensing, is perhaps the most difficult area as no licensing body would wish to be drawn into making statements and setting rules which allow software and computer systems to be generated to that set of rules, which the licensing body would subsequently have to abide by.

One aspect of the system not considered above is that most systems are built up from many components thus the project results must start by considering a single component and its justification. The combination of the components to form the system then occurs and the component justifications must be drawn together to give a basis for system justification. It is at this stage of the project that the major task of justifying quantitative measures of system reliability in terms of failure per demand will be made.

3.   THE APPROACH

The approach that has been adopted to obtain the information for completing the project is to undertake the exercise of procuring a computer based system for a safety critical application. During the process of procurement, the software production, assessment and licensing issues will be addressed to produce the required data. Indeed the experiment will concentrate upon these three activities. The decision was made to procure a system containing four diverse channels to perform a safety function. The production of a four channel system enables four implementations of the same specification to be produced, justified and assessed. This allows direct comparison of software engineering methods, the procurement cost and the reliability claims for the four channels to be made. The assembly of the four channels into a system then allows diversity and system justification to be considered in detail. Data on common failure should become available during the course of the experiment assisting a view on diversity to be taken.

There are a number of key aspects in setting up the experiment if there is to be a reasonable chance of success in its execution and if the results are to have value. This inevitably leads to the project moving through the traditional set of phases of an experiment. First, there is the set-up or planning phase in which the experiment is devised and means of gathering data established. Second, the experiment is executed and the data is collected. Finally in the third phase of the experiment the data and any other information collected during the experiment are analysed to produce the conclusions.

The points to be addressed in the planning stage include:

    selection of an example application
    selection of production methods
    selection of means of assessment
    selection of framework (rule) for licensing.

In addition it is necessary to:

    identify a means of acceptance testing
    identify a means of collecting data on personnel
    identify a means of collecting data on process
    identify a means of collecting data on product.

While the first four items are variables and while significant are not fundamental to the programme. The second four items are significant and results in DARTS, like many other advanced software projects, having a heavy metric content and inevitably becoming involved with international

(ESPRIT) and national metrics projects.

The second phase of the project sees the work concentrate on the implementation and assessment exercises. To a great extent this phase of the project requires little central guidance as the partners will be working in their domain of expertise. However attention must be given to ensuring the process data is collected during this phase of the project. In addition the project team must ensure that the implementors remain isolated from one another and from the assessors in order to ensure diverse implementation.

The third phase of the project is the analysis phase during which the main benefits are to be derived. The development of the analysis process will take place in parallel with the implementation and assessment. The early implementation and assessment reports being used as a starting point for the start of the analysis.

The DARTS experiment has in fact a key additional element. The discussion above has worked from the premiss of a single pass through the cycle. It is often the case that a change to the requirements is made during the course of procurement or once the system is in service. In these circumstances the issue of maintenance arises and the ability of the production process to accommodate such activity. The production exercise will therefore include a specification change that must be implemented following completion of the customer acceptance tests. This will test the ability of the methods to accommodate such change and examine the code maintainability, a non-functional but important requirement.

4.   THE TARGET SAFETY SYSTEM

The project partners generated five proposals for the target safety system of the DARTS experiment. These proposals were for safety systems based upon:

    Railway Signalling;
    Pressurised Water Reactor Protection System;
    Steam Generating Heavy Water Reactor Protection System;
    Ultra High Voltage Distribution Network Protection;
    Aircraft Conflicts Alert System.

Of these the Steam Generating Heavy Water Reactor example consisted of five elements:

    Neutron Flux Power Protection;
    Logarithmic Amplifier and Period Meter Protection;
    Primary Containment Pressure Protection;
    Steam Drum Pressure Protection;
    Steam Drum Level Protection.

In order to select the DARTS example each of the five proposals were considered against a set of features and properties considered to be important in the context of safety critical systems. The identified features are listed below.

    Time Dependence
    Sequencing
    Concurrence
    Arithmetic Processing
    Logic Processing
    Numerical Precision
    Complexity
    Response Time
    Number of Inputs
    Throughput of Input Data
    Number of Outputs
    Throughput of Output Data
    History/Memory Effects
    Human/External Intervention
    Security

FEATURES OF EXAMPLES

	Railway	PWR	Power	Power Doubling Time	Containment Pressure	Steam Drum Pressure	Steam Drum Level	High Voltage Protection	Conflicts Alert
Time Dependence	L X	H √	M √	M √	L √	M √	L X	H √	H √
Sequencing	H √	L X	L X	L X	L X	L X	L X	M √	L X
Arithmetic Processing	L X	H √	M √	M √	L √	L √	L √	H √	H √
Logic Processing	H √	L X	L X	L X	L X	L X	L X	M √	L X
Numerical Precision	L X	H √	H √	H √	H √	H √	H √	H √	M √
Complexity	H √	H √	M	M	L	L	L	M √	H √
Response Time	L X	M √	H √	M √	M √	H √	H √	H √	H √
Number of Inputs	H	M	L √	L √	L √	L √	L √	H √	H √
Throughput of Inputs		H √	M √	L √	L √	L √	L √	H √	H √
Number of Outputs	H	M √	M √	M √	M √	M √	M √	L X	L X
Throughput of Outputs		M √	L √	L √	L √	L √	L √	L X	L X
History/Memory	X	H √	M √	H √	M √	M √	L X	M √	H
Human Interaction	H X	L X	H √	L X	L X	L X	L X	L X	L X
Security	X	X	X	X	X	X	X	X	X
Minimum Size	√	√	(———————— √ ————————)					√	√
Maximum Size	√	H X	(———————— √ ————————)					H	H X

H High
M Medium
L Low

√ Appropriate
X Inappropriate

The size of the code had also to be considered particularly in an experiment such as DARTS.

The selection was made by conducting a marking exercise for each proposal. During the course of the exercise care was taken to ensure that any feature of the system that did not appear in the list but emerged as possibly being relevant was identified and included. The results of the marking exercise are given above.

The table shows that all of the examples display many of the features of a good example but that none of the examples is clearly better than the others. The selection process was therefore achieved by a process of elimination. The railway example is thus the first to be eliminated as it contains no time dependence or arithmetic processing and leaves much to be desired in terms of data throughput. Detailed discussion of the Pressurised Water Reactor example and the aircraft conflicts alert function example led to them being eliminated on grounds of size. It was considered that they are probably too big to be undertaken within the project budget and thus attempting them provided a major risk to successful completion of the project as neither could easily be descoped in the event of difficulty. The very high voltage network system example can clearly be tailored to the correct size however much of the control theory which forms a central part of the example would have to be excluded. This leaves the five functions from the SGHWR protection system example, these five functions are individually too small to be an example system but when collected together are too big. A number of the five functions can be combined to produce an example of appropriate size. Further adopting such an approach enables the different features identified as being important to be targeted in each of the systems. Further, should it prove necessary the requirement specification could be descoped should severe difficulties be encountered in any or all of the approaches to system production. The availability of this option will ensure that the project will produce the required information.

The example selected, taking into account problem size and possible risk in completing the experiment, lead to the selection of part of the SGHWR example. This requires trip action due to three separate measurements, these are:

- neutron power, top stop and rate of change, with an option on introducing an operator set level trip;

- steam drum pressure, top stop and rate of change;

- steam drum level, maxima and minima.

Each plant signal is supplied by three instruments in all cases thus the instrument signals must be checked for consistency and then processed through a trip algorithm. The software thus contains elements of both arithmetic and logic processing that might be found in a typical safety system and will be suitable for testing the software production methods. There is however little time dependency in examples as such, but should it be considered appropriate additional time effects can be introduced. This would be achieved by making it a requirement of the pressure or power trip processing that a compensation be made for the effect of the instrument damping which is introduced for noise rejection. The system would thus require that the true level and rate of change are compared with the trip levels.

5. ASSIGNMENT OF METHODOLOGIES

In setting up the project one of the primary questions that arose was which is the most cost effective way of producing, assessing and licensing safety critical software and procuring a safety critical computer system for operation? The emphasis is on the whole procurement cycle not just the individual activities of production assessment and licensing, as clearly the cost of one activity can be reduced with consequent increases in cost of the other activities. This can be illustrated by comparing the alternatives of adopting a strategy in which production costs are kept low but results in high justification and assessment costs with the approach that accepts high production cost but consequently has low justification and assessment costs.

The questions that arise from this apparently simple question of cost effectiveness and that are to be addressed by the DARTS experiment include:

Which type of method for production, assessment, licensing leads to the procurement of safety

ALLOCATED METHODOLOGIES

	CHANNEL 1	CHANNEL 2	CHANNEL 3	CHANNEL 4
METHOD	FORMAL	FORMAL	TRADITIONAL	TRADITIONAL
HARDWARE	INTEL 8088	TRANSPUTER	MOTOROLA 68020	MOTOROLA 68020
SPECIFICATION				
METHOD	VIENNA DEVELOPMENT METHOD	MODAL ACTION LOGIC	IDEF.O & STATE MACHINES	YOURDON CONSTANTINE
TOOL	SPECBOX & MURAL	FOREST	ASA	TEAMWORK SA TEAMWORK RT TEAMWORK D DREPORT
ENVIRONMENT	PC/MS-DOS	SUN3 & 4/UNIX	SUN4 / UNIX	μVAX 3400/VMS VAX STATION 3100/VMS
DESIGN				
METHOD	JACKSON & VDM	MODAL ACTION LOGIC	HIERARCHICAL OBJECT ORIENTED DESIGN	YOURDON CONSTANTINE
TOOL ENVIRONMENT	PDF + MURAL μVAX 3100/VMS + SUN 3/UNIX	FOREST SUN 4 / UNIX	STOOD SUN 4 / UNIX	TEAMWORK SD μVAX 3400/VMS VAX STATION 3100/VMS
CODE				
LANGUAGE TOOL	PASCAL COMPILER & DEBUGGER VAX SET	OCCAM	ADA COMPILER & DEBUGGER	C CLANDS VAX SET
ENVIRONMENT	μVAX 3100/VMS	PC/MSDOS	SUN 3 / UNIX	μVAX 3400/VMS VAX STATION 3100/VMS
ANALYSIS TRANSFER	PVL SPADE LDRA TESTBED MICE	NOT REQUIRED	VERILOG LOGISCOPE ADA CROSS COMPILER	MICE
TESTING HOST	DYNAMIC ANALYSIS + PERFORMANCE & COVERAGE ANALYSER	NONE	DYNAMIC ANALYSIS	PERFORMANCE & COVERAGE ANALYSER
TOOL	LDRA TESTBED VAX SET		LOGISCOPE	VAX SET
TESTING TARGET	MICE	TRANSPUTER DEVELOPMENT SYSTEM		MICE

critical software at least cost?

What are the relative costs of approaches using formal methods and traditional methods? Is the current presumption that the latter are cheaper than the former correct?

Is there, or should there be, additional confidence in formally produced software, and if so is the additional confidence obtained in a cost effective manner?

What is the nature of the problems faced by an assessor when confronted by a system to assess and is this affected by the use of formal methods and tools which require a very high level of specialist knowledge?

This discussion leads rapidly to the identification of a number of important issues. These include:

The use of formal methods

The interactions of the parties involved in procurement lifecycle

The use of tools and benefits of tool support for production and assessment activities

The interaction of the members of an implementation team

The methods of justifying the high reliability requirements of the system

In an exercise of this nature where there are a large number of variables but only a limited number of experiments possible there is the inevitable conflict between the desire to exercise as many of the variables as possible and retaining sufficient control so that a useful comparison exercise can be undertaken. One consequence of the large number of methods is that not all the questions will be completely answered. Thus the selections made must contain only those aspects of the highest industrial relevance. The methods chosen for the four production teams and from which the assessment and licensing processes will follow are given above.

The allocation of the methodologies was completed to give two traditional method channels, one established formal method channel and a second formal channel that contains many novel features. This weights coverage of methodologies towards traditional methods as they are currently of most practical interest.

The case for formal methods has still to be proved and the necessary skill and tool base built up. The inclusion of the HOOD, ADA approach is regarded as being highly significant as it appears to be a route that is gaining much favour in industry and with licensors. This is balanced by the traditional approach through structured methods as represented by the TEAMWORK tool and the formal methods approach through VDM. The wild card FOREST OCCAM route is very much the research route that is pushing the frontiers of current technology. The inclusion of this channel recognises the very rapid advances currently being made with transputer technology and the moves to include them in process control systems.

The balance of the project is maintained in a similar fashion with respect to the assignment of programming languages but here the shift is to a more structured approach with PASCAL and ADA; the traditional approach being represented by C. This balance has been chosen to reflect the current industrial view that the use of strongly typed programming languages is preferred for safety critical systems. The omission of ASSEMBLER is a disappointment but is necessary given the limited number of channels. The use of OCCAM as the novel aspect should generate some interesting discussion and results on the topic of languages for safety critical systems.

The testing strategies and the links to the use of tools by producers, assessors and licensors covers the range of: no analysis tool being available for channel two, through to possibility of full analysis and tool support for channels one and three. The balance towards the use of testing tools again reflects the perception of the current view of both industry and licensors that such tool support is desirable. Tool use is also increasingly favoured by producers as it automates much of the testing, with good quality tools producing test logs and reports as part of the package, helping quality assurance and reducing the amount of skilled effort required for testing.

The approach to assessment and licensing will be retrospective in all cases. This is necessary to prevent the assessors becoming part of the development process and possibly prejudicing the experiment. This decision was the centre of considerable debate as it is the view that non-retrospective assessment has the potential of being a more efficient process than retrospective assessment. However the benefits available from such an approach cannot be realised in this experiment. There must be a major question if these benefits can ever be realised as should the assessor become involved with the process then his independence must be compromised. The licensing and assessment is discussed further below.

6. FRAMEWORK FOR LICENSING AND ASSESSMENT

Many development standards exist for safety critical systems and software for example IEC 880 and RTCA 178A/EUROCAE ED-12A standards and further work is in progress with standards being developed by IEC SC65 WG9 and WG10. In addition, the basic principles standards for assuring quality in such systems are well established , see for example ISO 9000 and BS5882 and specific guidance is available for quality control and software engineering practices for example DOD-STD-W167 and ESA-PSS-05-0.

However the situation is quite different for the licensing process as there is little guidance or consensus on the licensing process for software-based systems. A major study commissioned by the EEC on licensing in the nuclear industry, report EUR-11147EN, 1987, showed that there was little commonality of approach in the EEC member countries. A further study commissioned by the British Department of Trade and Industry, BSI/IEE Report on Software in Safety-related Systems, showed that there were considerable differences in practice between different industries and countries in their approach to the regulation of software-based systems. The relationship between the regulators and the developers ranged from extremely detailed independent assessments to a more 'hands off' approach where the licensing is based on information provided by the developer.

In order to address this problem, the DARTS project partners have developed a framework document for the licensing of software-based systems. It has been based on their combined experience of the licensing process and represents their consensus view of the 'best practice' licensing approach. In the experience of the project partners, many of the problems encountered in the licensing process arise at the points of interaction of parties with different interests and responsibilities. By providing a generic, structured framework to ensure that the relationships and responsibilities of the regulator, the customer and developer are well-defined conflicts should be avoidable. The framework also requires that the development, justification and licensing activities should be planned and scheduled together so that problem areas are rapidly detected and licensing delays are minimised.

We hope to investigate and validate this approach by applying it to the case of the DARTS project and where possible, by promoting its usage on an experimental basis in actual industrial projects.

The basic objectives of the Regulator are to ensure that:

- the correct safety functions have been specified;

- these functions have been correctly implemented;

- safety will continue to be maintained in operational life.

In order to meet these overall objectives, the licensing framework uses the following basic philosophy:

- Integration with System Safety. The safety assurance activities of a PES should form part of a coherent safety assessment and licensing approach for the whole system. The functional behaviour and other requirements for a PES should link back to the safety case for the plant.

- Licensing Based on a Combination of Assessments. No single method will give complete assurance of a PES safety function, especially in software systems which contain a high degree of internal complexity. The basic approach is to base assurance on assessments of the:

  . the development processes (management, personnel, methods)
  . the product

- Non-prescriptive. The framework does not enforce a particular implementation approach. Any approach can be accepted provided it can be demonstrated to satisfy the overall licensing objectives.

- No Surprises. The basis for assessment should be made clear and agreed from the outset of the project.

- Least Effort. Ideally (from the assessors viewpoint) the Customer and supplier should provide most of the justification for the system and this information should be compatible with the licensing approach.

- Continuous Assessment. Problems and unacceptable practices are much easier to detect and correct if they are found early. Involvement of the Regulators from project initiation should minimise rework, delays and result in a more assessable system. Should continuous assessment be adopted, the assessor must ensure that he does not compromise his independent position.

The implementation of the ideals above leads directly to the production of a generic framework which can be customised for a sector specific application. In the case of DARTS this was done for the nuclear sector. However it is emphasised that this customisation did not need to take into account nuclear issues as such, the customisation was for computer issues only.

7. ACCEPTANCE TESTS

Acceptance testing still appears to form the basis for decision making associated with the decisions to accept, pay for or license the system. The discussion of acceptance tests resulted in three phases of tests being identified. The actual testing performed at each phase is very similar, the objective being to show that the system conforms to the requirements specification and is thus strongly biased towards functional testing. However as the tests are undertaken for different reasons, they place emphasis on different aspects of testing.

The three phases of acceptance tests identified are:

- the _factory_ acceptance testing phase, where the implementor intends to ensure that the product corresponds to his/her own interpretation of the specified task before releasing it to the customer

- the _customer_ acceptance testing phase, where the implementor has to demonstrate that the product corresponds to the customer requirements. The product is often paid for against the results of these tests

- the _licensor_ acceptance testing, this is the phase where the assessor and licensor parties will determine the acceptability of the products by means of comparative evaluations.

The generation of input data for testing is often a problem. Simulation techniques are coming to the fore. In the case of the DARTS project the full SGHWR reactor simulator that is available at Winfrith will be used to generate basic test data. The experimental environment will however allow the simulator data to be perturbed and random test data to be applied. It is expected that a very comprehensive range of realistic test data including transients will be used during the experiment.

8. DATA COLLECTION

There are two primary types of data collection that are undertaken during the course of the DARTS project. The first type of data collected is the effort data, the second type of data is error data, both of which are collected during the project. In addition a considerable amount of data will become available from the quality records of the producers and assessors. This data will for example include the test records.

The background of two of the partners and their involvement with work on electronic databases to store software metrics information lead to the decision to collect and record both the effort and the error data through electronic forms. Both special forms and the supporting data repository have been coded and are in use to gather data. There is a considerable amount of software product metric data that will be generated by both the producers and assessors, this again needs collecting and collating. This will be undertaken using the product metrics database based upon that produced for the SCOPE project.

9. CURRENT STATUS

The project is currently midway through the implementation exercise. The producer teams have generated their quality plans and forwarded them to the assessors who have used them to develop up their assessment plans. These quality and assessment plans form the basis of the programme against which the work is to be carried out. The producers have moved well down the road of producing their specifications and in a number of cases moved on into the design phase.

Data collection has proceeded quite well, considerable attention has been given to ensuring that the effort data records are maintained in good order. Less attention has been given to collection of error data, this has been possible because error data should in principle be available from project quality records.

The main thrust of the forward programme is the move to completing the four channels and moving them to Winfrith for integration into the safety system and test harness for final system testing and data gathering. A common PC based factory test harness is being produced for channel testing, ie factory and customer acceptance testing prior to channel shipment to the UK. It is believed that this will ensure that no unexpected problems occur on system integration.

10. ACKNOWLEDGEMENT

The work reported here is compiled from the partners in the DARTS project. The efforts and contributions of staff at Ceselsa, Electricité de France, Gesellschaft fuer Reaktorsicherheit, Nuclear Electric and AEA Technology are acknowledged along with the financial support of the project by these companies and the CEC.

A COMPUTER SYSTEM APPLICATION TO IMPROVE NUCLEAR PLANT AVAILABILITY

A. De Martinis* and A. Pasquini**

*C.I.T.E.C. S.pA., via A. Farnese 3, 00192 Roma, Italy
**ENEA, via V. Brancati 48, 00144 Roma, Italy

Abstract - Safety criteria in nuclear power plant require that the process control is based on both a supervision and process control system as well as a protection system; the first acquires and processes the critical process parameters and consequently controls its state, whereas the latter stops such process should potentially dangerous conditions occur. However, protection system actuations will involve various serious consequences ranging from the introduction of mechanical stress into the plant systems, to very long out of operation times with economical overall losses as a result. The interventions of the protection system are not always foreseeable through the control system since it is mainly dedicated to the optimization of the process. This paper describes the architecture, the functions and the verification and validation plan of a computerized system that is specially designed to prevent protection system actuations. This system alarm the operators when conditions requiring plant SCRAM are going to occur and gives them the necessary indications to bring the plant back to normal operational conditions. This system has been installed on a fast nuclear reactor used in an Italian nuclear research center.

Keywords - Software development; computer control; program testing; nuclear plant availability; verification and validation plan.

I - INTRODUCTION

Safety criteria in nuclear powerplants require a physical and functional separation between control and protection system. The first one controls the nuclear process and usually performs process stabilization, core nuclear power control and process parameter optimization to maintain the plant in safe operation.

These functions are more and more frequently assigned to control process computer systems (Botts, 1985) that, in some cases, directly give actuation commands to control rod drive, recirculation pumps, etc. (Gilbert, 1985) while in other cases provides only information to plant staff without giving actuation commands (Fahley, 1985).

The protection system senses process conditions and generates signals for reactor trip and for initiating corrective actions if process conditions exceed specified limits. The possible corrective actions range from containment isolation to emergency core and containment cooling and the correct selection of their sequence is a complex operator task due to the recent grow in size and complexity of the nuclear power plants. For this reason computer systems are becoming more widely used even for this kind of application (Bacher, 1985; Bagnasco, 1985). In addition computers are used to support plant staff during abnormal conditions, by performing the diagnostics functions, that is, providing alarm selection and presentation, process component status display and suggesting the right sequence of actions to limit failure effects and mitigate their consequences. A classical example of such systems is the Emergency Information System (EIS) described in (Ceriati, 1985).

Due to the mentioned safety criteria the protection system does not contribute to normal plant operations and only intervenes under abnormal circumstances. However its actuation involves expensive consequences for the plant. The sudden change in the status plant caused by scrams induces abnormal transients that could cause other failures of the plant systems. Further the mechanical stress introduced into the plant systems can significantly reduce their operating life. At last, a long time is required to return to normal operating conditions since the core is "poisoned" by the fission products (e. g. xenon, argon, krypton) and the plant can only start up after a certain period of time with the consequent economical losses for the utility of the plant.

For these reasons conditions requiring protection system actuation should be carefully avoided and several efforts have been devoted to this aim. But these conditions can not always be predict using the normal control system since it is mainly dedicated to the optimization of the process. It would be desirable to be advised as soon as process disturbance starts, in order to have warnings sufficiently early to avoid taking drastic countermeasure in restoring normal plant conditions. But if the warning limits of the control system are too close to the desired operating point that would cause too many messages during dynamic plant operations (Ness, 1986). On the other hand there is a general agreement that plant safety and availability can be enhanced by providing the operator with more operational support, if that can be done without overloading him with unnecessary information (Berg, 1985; Lupton, 1989).

The system described in these paper is specifically devoted to prevent protection system actuation through an early detection of possible process abnormal conditions. It is installed in parallel with the conventional control system in a fast nuclear reactor of an Italian research center and it emphasizes such conditions giving to the operators information and suggestions to early restore normal plant conditions.

Section II describes the functions and the architecture of the system and shows the approach used to prevent conditions requiring protection system actuations. Section III describes the verification and validation plan adopted during development. Finally section IV outlines the first results and conclusions that can be drawn from its application.

II- SYSTEM DESCRIPTION

The reactor, TArature PIle pow r zeRO(TAPIRO), is an high enriched Uranium-235 fast neutron source facility located at the ENEA CRE, Casaccia, near Rome.

This reactor is characterized by a neutronic flow (4.E + 12 neutrons/cm**2 sec) that is exceptional if compared to the dimensions of the plant (Angelone,1987). Its registered average number of turnoff requests is considerably higher than demands, i.e. the condition when one or more system variables assume values which cannot guarantee a correct functioning of the plant. This is due to various reasons:

a) The reactor is used for fast reactor shielding experiments, blanket studies, fission yields evaluation from hard fast reactor core spectrum to degraded blanket/reflector spectra, as well as for in   irradiation; consequently the reactor is not working continuously, and it has frequent startups, power raising and turnoffs. The considerable amount of transients which this kind of functioning involves, can easily cause protection system actuations.

b) The nature of the research experiences carried out often requires a "step-by-step" power progress .

c) System SCIN, Strumentazione di Controllo Impianti Nucleari, has been designed according to a Fail-safe principle without taking into account the plant availability.

Due to these characteristics, Reactor startup is a complex phase requiring careful operator attentions. After start up, when the plant has reached the prefixed amount of power, it is necessary to maintain this value; in other words, that the ongoing nuclear reaction remains within the established limits, without variations. This is the only occasion to use the Reactor for irradiation experiences or nuclear physics studies. It is possible that one or more parameters exceed the programmed alarm thresholds and cause a SCIN actuation, which interrupts the nuclear reaction causing a SCRAM. For these reasons operators have been supported with another system, in addition to the control system, especially designed to prevent conditions requiring a protection system actuations. In this manner it has been obtained a drastic reduction of SCRAM occurrences, reducing the possibility that SCRAM occurs.

Particularly serious problems with regard to space and various practical problems have not permitted us to insert new sensors in the plant and consequently it has proven necessary to take the major part of the information from the already existing protection instruments.

This system performs the following functions:

a) Integrates the information supplied by the SCIN system and displays the dynamic evolution of several variables in different ways.

b) Makes correlations between the variables under observation and selects the more significant subset of them. In other words the pre protection system selects variables related with those under observation; it controls and presents them to the operator only if they reach critical values.

This spares the operator from taking under control the whole set of variables and leave him to follow only the subset of variables which may cause SCRAM or in any case be considered as priority. Furthermore, the system automatically updates the correlations between the variables on the basis of the most recent acquisitions; in this manner, the attention is drawn to the variables with more tendency to exceed the prefixed limits in respect of those with a stable behavior.

c) Allows the operator to follow the evolution of the most critic variables to be kept under monitoring; various types of formats are available: numeric tables, as offered by the instrumentation of the control system; graphic form,with zooming capability in order to analyze the minimum variation of amplitude.

d) Predicts possible variable evolutions. It computes the current data and analyzes the most significant historical data in order to suggest the possible evolution for particularly critical variables to the operator. The capabilities of a normal real time data processing system are not sufficient for these activities and consequently, on the basis of similar experiences (Kaemmerer, 1985), we decided to use an expert system able to analyze the data of the historical database of the plant.

e) logs the whole set of plant variables during SCRAMS or during prefixed intervals of time as defined during start-up.

The pre-protection system hardware architecture is shown in fig. 1. It is composed of a Digital Microvax II mini computer, a Real Time Data Acquisition Subsystem (RTP), a Fail Safe Interface Unit to SCIN (FSU) and several Graphic Workstations (GWS), which are connected to the central unit through a LAN Ethernet. These workstations are located in the console room and in the reactor room, but other stations are directly used for monitoring particular experiences in nuclear and biological physics; in the future other workstations are planned for retrospective studies of the historical data acquired.

The software architecture is shown in fig. 2. It is based on several modules, strictly correlated; the task monitor (ACQMONIT), controls the synchronization of the processes constituting the system pool and their priorities.

The duty cycle of the data acquisition system is one second; the first 300 msec are used by the RTP subsystem to acquire the signals originating from the FSU unit, under the control of a process (ACQRTP) and at the maximum priority, and in the remaining 700 msec the acquired data are transmitted through Q- BUS to the Vax , to be processed.

The task process (PRELAB) transforms data into engineeristic units and compares them with the specific thresholds; on the basis of these comparisons and of parametrical standards established at initialization time, the proces sed variables may be subject to further processes and could assume the proper priority codes for the next steps. The task PRELAB takes into account the historical trend of the controlled variables to follow their evolutions and to compare the foreseen trend with thresholds.

The module COMGR dispatches data between historical Data Base, the display terminals GWS and the service routines; it also manages the human interface functions by selecting the variables to be presented and the methods of their presentation.

The supervisor task ACQMONIT also manages the requests submitted to the central system by the various users working on the GWS, e.g. new aggregations between variables or scale changes and a network mailbox available for every node of the system to deposit requests transmitted by the GWS to the central unit.

III - VERIFICATION AND VALIDATION

The system we are dealing with is not directly safety related since it does not perform protection functions. Further it provides only information and suggestions to plant staff and does not perform directly actuation commands. Then its possible failure does not lead to significant degradation of plant performances.

Nevertheless some safety requirements have been taken into account during its development: separation must be assured between this system and the protection system in such a way that no failure of any component of this system or of common components can affect the functionality of the protection system; the system usability and availability should avoid that the operator wrongly or ill informed (by the system) could lead the plant in a potentially dangerous state requiring the protection system actuation. Compliance with similar requirements was used as an assessment criteria for systems with similar functions (De Agostino, 1985; Di Marco, 1988). Further, even if not safety related, the system functions can significantly reduce the potential dangerous states and the economical losses associated with scrams.

For this reason fault avoidance techniques adopted during development have been completed with an accurate verification and validation plan to allow for early detection and correction of faults and to ensure compliance with the requirements (EWICS, 1988). The verification and validation plan has been developed during the software development phase usually called Requirement definition and Project organization.

The plan set forth the approach to be followed for conducting the verification and validation activities (EWICS,1988) and in particular it defines: scope and objectives of the verification and validation plan; documentation to be produced; teams to be involved in the verification and validation activities; related responsibilities; techniques to be used during each development phase and related tools. Several of these techniques are available (EWICS, 1990) and their selection is one of the most critical part of the plan. Table 1 shows for each development phase: the related product subjected to an activity of verification and validation; the adopted technique and the reasons for its selection.

Notes to tab. 1:

(1) The system is not directly safety related and does not perform direct actuation commands. For this reason formal methods and program proving have been considered too expansive even if applied to the most critical part of the system and Inspection (Freedman, 1982) has been considered sufficient for a system of this criticality and complexity.

(2) The system customer has a large knowledge and experience with computer systems and he produced very precise, well defined and definitive requirements. For this reason and for the same economical reasons of the previous point no prototyping or symbolic execution have been adopted for this verification phase.

(3) Modules were initially tested with the aim of executing every statement and every branch at least once. Then, after integration, the whole system was functionally tested to check the fulfillment of all specified requirements. The coverage obtained with the latter approach was analyzed and a new structural testing was performed with the aim of covering all the statements non executed after integration

(4) A less complete functional testing (only the most simple error conditions were considered) was performed during the validation phase. Then the system was executed in the final version using a system driver and an off- line plant simulator. During this phase test were selected randomly but in accord with the operational demand profile (Myers, 1979).

IV - CONCLUSIONS

The system described has been under testing for more than one year and has meeting the re quirements very satisfactorily; however, since the system has not yet been completely implemented, it is not yet possible to give a complete evaluation of its performances.

The principal aim of the system was a significant decrease of the number of unintentional protection system actuations, in order to increase plant availability. Although the plant still suffers from minor problems which need adjustments, we can definitely state that the number of SCRAM has been considerably reduced. Furthermore, operators and plant users deeply appreciate the logging capacity of the pre- protection system, which is able to present particularly important data acquisition intervals for diagnostic purposes. Plant logging was also used to create a historical data bank, on the basis of which it will be possible to obtain a further SCRAM reduction and to create new correlations between the plant variables.

However we wish to point out that the large number of variables shown with the present display system may in some cases overload the operators or disturb their activities. We are presently planning to overcome this drawback by using A.I. techniques, both for the selection of the most significant variables to be presented and for the selection of the presentation techniques.

V - REFERENCES

Angelone, M., P. Moioli, and G. Rosi (1987). The High enriched Uranium 235 copper reflected TAPIRO reactor. VEL-MEP ENEA Int. report, Roma, Italy.

Bacher, P., and G. Guesnier (1985). French Approach in the Use of Computers for Monitoring and Control of the Nuclear Power Plants. Proc. of the Int. Topical Meeting in Computer Applications for Nuclear Power Plant Operation and Control, Pasco, USA, pp. 32 - 38.

Bagnasco, S., F. Manzo, and F. Piazza (1985). Requirements and Design for a Distributed Computerized System for Safety and Control Applications. Proc. SAFECOMP '85, Como, ITALY, pp. 85 -93.

Berg, O., O. Evjen, U. S. Jorgensen, J. Kvalem, and I. Leikkonen (1985). Early Fault Detection Using Process Models and Improved Presentation Techniques. OECD Halden Reactor Project, HWR 141, Norway.

Botts, W. V. Jr. (1985). Computers and Nuclear Power -- Past, Present, and Future. Proc. of the Int. Topical Meeting in Computer Applications for Nuclear Power Plant Operation and Control, Pasco, USA, pp. 817 - 820.

Ceriati, F., G. Guelfi, and F. Gangemi (1985). Computer Based System Application for Plant Monitoring in Italian BWR - 6 Alto Lazio. Proc. Computer Applications for Nuclear Power Plant Operation and Control Int. Topical Meeting, Pasco, USA, pp. 538 - 542.

De Agostino, E., V. Silvani, and F. Zambardi (1985). Design Criteria Evaluation for Computerized Safety Systems. Proc. Computer Applications for Nuclear Power Plant Operation and Control Int. Topical Meeting, Pasco, USA, pp. 337 - 443.

Di Marco, G., and A. Pasquini (1988). Criteri di licensing per sistemi computerizzati in impianti nucleari. Sicurezza e Protezione, no. 47, Roma, Italy.

European Workshop on Industrial Computer Systems (EWICS) T. C. 7 (1988). Dependability of Critical Computer Systems I. F. Redmill (Ed.). Elsevier Applied Science, London.

European Workshop on Industrial Computer Systems (EWICS) T. C. 7 (1990). Dependability of Critical Computer Systems III. P. Bishop (Ed.). Elsevier Applied Science, London.

Fahley, J. M., and J. K. Bigelow (1985). Computing and Use of Digital Computers at the Diablo Canyon Power Plant. Proc. of the Int. Topical Meeting in Computer Applications for Nuclear Power Plant Operation and Control, Pasco, USA, pp. 55 - 61.

Freedman, D. P., and G. M. Weinberg (1982). Handbook of Walkthroughs, Inspections and Technical Reviews: Evaluating Programs, Projects and Products. Third ed. Little, Brown and Co., Boston, USA.

Gilbert, R. S. (1985). Control and Safety Computers in CANDU Power Stations. IAEA Bulletin, Autumn 1985, pp. 7 - 12.

Kaemmerer, W. F., and P. D. Christopherson (1985). Using Process Models with Expert Systems to Aid Process Control Operators. Proc. of the 1985 American Control Conference. ACS, Boston, USA.

Lupton, L. R., J. J. Lipsett, and R. R. Shah (1989). A Framework for Operator Support Systems for Candu. Proc. of the IAEA/IWG Specialists' Meeting on Artificial Intelligence in Nuclear Power Plants. VTT, Helsinki, Finland.

Myers, G. (1979). The Art of Software Testing. Wiley and Sons, NY, USA.

Ness, E., O. Berg, and A. Sorenssen (1986). Early Detection and Diagnosis of Plant Anomalies Using Parallel Simulation and Knowledge Engineering Techniques. Proc. of the Int. Topical Meeting on Operability of Nuclear Power Systems in Normal and Adverse Environment. American Nuclear Society, Albuquerque, New Mexico.

Development phase	Development product subject to V & V activity	V & V techniques adopted					
		Inspection	Program analysis	Functional testing	Structural testing (decision coverage)	Structural testing (statement coverage)	Random testing (operative profile)
Requirements definition	Requirements specification	X (1) (2)					
Design	Design specification	X (2)					
	Preliminary program documentation		X				
Coding	Program document.		X				
	Source code				X (3)		
Testing and Integration	Integrated code			X (3)		X (3)	
	User documentation	X					
Validation	Final system			X (4)			X (4)

TABLE 1 - Techniques of the verification and validation plan

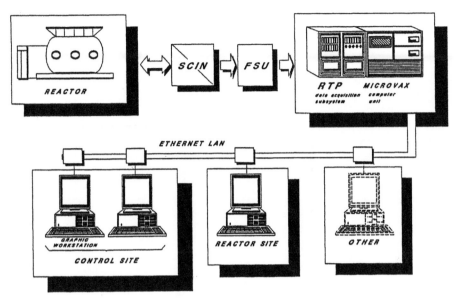

Fig. 1. Hardware Architecture

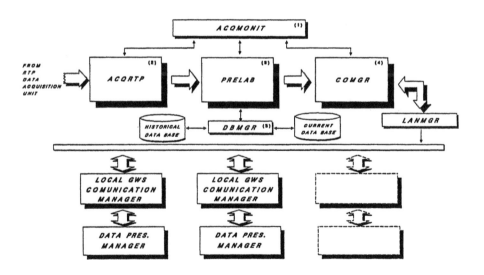

Fig. 2. Software Architecture

Fig. 1 Hardware Architecture

Fig. 2 Software Architecture

A SOFTWARE TOOL FOR FAULT-TOLERANT SYSTEMS IN THE OPERATIONAL PHASE

K. Sismail

Faculté des Sciences. 2 Rue de la Houssinière, 44072 Nantes, France
and
IRISA, Campus de Beaulieu, 35042 Rennes, France

abstract. A software tool for predicting reliability and safety measures
for computer or dynamical systems in the operational phase is presented.
The considered models are markovian or semi-markovian. The background
material which allows such predictions is based on new filtering formulas.

Key-words. Fault-tolerance; filtering; markovian models; operational phase;
reliability measures; safety; semi-markovian models.

INTRODUCTION

In view of fault-tolerance in reliable fonctioning of computer or industrial systems, a system may go through as many as 10 stages in response to a failure as it is pointed in [5]. Designing a system involves the selection of a coordinated response that combines some or all of these steps :
* fault confinement, limiting the spread of fault effects to one area of the system;
* fault detection;
* fault masking, hiding the effects of failures;
* retry, a second attempt at an operation is often successful, particularly in the case of a transient fault;
* diagnosis;
* reconfiguration, reconfiguring a component to isolate it;
* recovery, backing up system operation to the point prior to fault detection;
* restart from the recovery point;
* repair;
* reintegration, placing a repaired module back in the system after physical replacement of a component.

These steps are not instantaneous and some of them may take a long time. Therefore the user (or the entity) which controls the system behavior must be able to take decisions on the basis of the data collected at the time of implementation of the control, in order to respond to differents situations. If possible, the decisions have to be optimal in some sense. Such decisions can be for instance: 'continue functioning', 'increase the redundancy of some units without interrupting the service' or 'stop the system' to avoid severe risks (this is the case for a nuclear station, automatic steering, ...).

Hence, on line predicted reliability measures are an essential issue. They can be achieved by taking into account the past behavior or the observed history of

the system. A "history" consists of a complete or an incomplete information on the behavior of the system up to a time t, called the final observation time, at which the user has to take a decision. For the complete information case, all the "events" and the times at which they occur are known from the user. In the case of an incomplete information, only some events are known from the user. For example, he(she) knows that a system unit failed before the final observation time but does not know the exact instant of the failure occurence. Conversely, he(she) knows the exact instant of failure but not the nature of the faulty unit. And a decision should be taken in real time before finding the faulty unit.

Systems under study are general. A system unit is any electronical, electrical, mechanical or software item subject to failures (and in most of the cases to repairs). In the case of a software item and depending on the available data, modules or recovery blocks (used for fault-tolerance mechanism) may be considered as system units. Our work focuses on the computation of reliability measures, including safety, in the operational phase on the basis of various possible histories of the system. The considered models are markovian or semi-markovian, which are commonly used to model fault-tolerant systems. The mathematical background used is based on filtering formulas.

For the user, a reliability tool SOLAR is presented. A user oriented language is given. At the input level, the description of the system in terms of components characteristics, dependencies between components, minimum numbers of components under which the system is considered operational (respectively safe) and in terms of repair strategy and the history of the system up to the final observation time are entered. In the output, a reliability menu gives the reliability and safety measures with or without the utilization of past behavior. A plotting module

47

supports displaying reliability data versus time.

This work is organized as follows. We first introduce the different kinds of histories and we show how to compute the reliability and safety measures in the operational phase. Mathematical backgrounds used to perform such computations are given. New filtering formulas are proved. Then the software tool SOLAR and its user syntaxic language are presented. The ouputs which are reliability and safety measures in the operational phase are shown on a multiprocessor and on a dynamical system.

SYSTEM HISTORIES

A system history is the ordered sequence of overlapping steps outlined above. Fig. 1 shows an example of system behavior where only the fault detection, fault isolation and repair steps are considered. An event can be a failure, a repair, a fault detection, a fault isolation. In the sequel and for simplification, we restrict ourselves to histories which are only described by the events of the following kinds: failures and repairs of hardware or software units, and their occurence times.

The class of events, denoted by $\mathcal{E}$, is defined as follows. Let a system be constitued by L kinds or types of units. A type of unit is a unit which differs from the others by its physical nature or by its failure or repair rate. For example, let us consider a multiprocessor with two processors and three memory units connected to a bus. Assume first that the memories (respectively the processors) have the same failure and repair rates. Then the memory, the processor and the bus are types of units. Assume now that the failure and repair rates of the processors differ, in which case the two processors are distinguished by processor 1 and processor 2. Then the memory, processor 1 and processor 2 and the bus are types of unit. During the observation period, the user can observe failures or repairs. For each failure or a repair, he(she) may or may not identify which unit is concerned. Then, the set of possible events is

$\mathcal{E} = \{failure, repair, failure(unit\ of\ type\ l),$
$repair(unit\ of\ type\ l), 1 \leq l \leq L\}.$

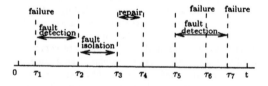

Fig. 1. Time order of events.

Let us denote by $\mathcal{H}_t$ the observed history of the system until time t and by $\tau_1, \tau_2, \tau_3, ..., \tau_n$ the ordered sequence of the instants at which the events occur. The instant τ_n is the last event occurence before the final observation time t. The general formulation of $\mathcal{H}_t$ is

$$\mathcal{H}_t = \{(\tau_i, e_i), i = 1, ..., n\} \qquad (1)$$

where $e_i \in \mathcal{E}$ is the event which occured at τ_i. Such a history is said to be complete if for each $1 \leq i \leq n$, τ_i is known from the user and $e_i \in \{failure(unit\ l), repair(unit\ l), 1 \leq l \leq L\}$ (i.e. the type of each failed or repaired unit is identified). Otherwise we say that the history is partial or incomplete. An example of an incomplete history is when the τ_i's are not observed, i.e.

$\mathcal{H}_t = \{e_i, i = 1, ..., n\}.$

Modelling

Through a user oriented language, the characteristics of the system are entered in SOLAR (see below for more details) which builds automatically the appropriate stochastic model describing the system. Such a model depends on the distributions of the time to failure or the repair time of the system units. We assume that these distributions are one of (or are approximated by one of) the following :

- the exponential distribution,

- the 2-hyperexponential distribution, denoted by $hyper(\lambda_1, \lambda_2, \alpha)$, for which the density writes
$$f(t) = \alpha\lambda_1 e^{-\lambda_1 t} + (1 - \alpha)\lambda_2 e^{-\lambda_2 t}, \quad t \geq 0$$
where $0 \leq \alpha \leq 1$.

- the generalized k-Erlang distribution, denoted by $Erlang(k, \beta, \mu)$, whose the density is given by
$$f(t) = \beta e^{-\mu t} + (1 - \beta)\mu^k \frac{t^{k-1}}{(k-1)!} e^{-\mu t}, \quad t \geq 0.$$
with $0 \leq \beta \leq 1$.

The choice of these phase-type distributions is motivated by the following reasons:

- to build models tractable by markovian methods (which are efficient to compute transient measures),

- to allow the approximation of general distributions according to a rule based on their coefficient of variation.

For system modelling, we assume that a probability space $(\Omega, \mathcal{F}, P)$ is given and we denote by X_t the system state at time t taking its values in an enumerable space E. In what follows, we distinguish the "user level" at which the user expresses an history in terms of events (failures or repairs) as in (1) from the "model level" where such a history is expressed in terms of the transitions of the stochastic process X_t modelling the system. We give below an example where we show how the two levels interact.

A Basic Example

Let a system be reduced to a simple unit subject to failures and repairs. At the user level, a history may be

$\mathcal{H}_t = \{\tau_1, failure, \tau_2, repair, \tau_3, failure, \tau_4, repair,$
$..., \tau_n, failure\}$

which here is a complete information (since there is only one unit). Let us now translate such an information at the model level, i.e. in terms of states of the system. We denote by $\mathcal{G}_t$ the history in terms of X_t induced by $\mathcal{H}_t$. We consider the following cases :

(i) the time to failure and the repair time of the unit are exponentially distributed. The states of X_t which correspond to a failure and a repair are respectively denoted by 1 and 2. Therefore

$$\mathcal{G}_t = \{\tau_1, X_{\tau_1} = 1, \tau_2, X_{\tau_2} = 2, \tau_3, X_{\tau_3} = 1, \tau_4, X_{\tau_4} = 2, ..., \tau_n \leq t < \tau_{n+1}\}$$

which is a complete information for X_t since all the visited states are known.

(ii) the time to failure and the repair time of the unit are respectively $hyper(\lambda_1, \lambda_2, \alpha)$ and $hyper(\mu_1, \mu_2, \beta)$ distributed. The appropriate markovian model describing the system is the one on Fig. 2.

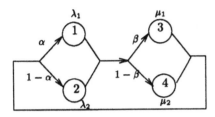

Fig. 2. The 2-hyperexponential failure/repair case.

Let $A_1 = \{1, 2\}$ and $A_2 = \{3, 4\}$. It yields

$$\mathcal{G}_t = \{\tau_1, X_{\tau_1} \in A_1, \tau_2, X_{\tau_2} \in A_2, \tau_3, X_{\tau_3} \in A_1, \tau_4, X_{\tau_4} \in A_2, ..., \tau_n \leq t < \tau_{n+1}\}$$

$\mathcal{G}_t$ is an incomplete information for X_t since only specific groups of states (not the exact states) are known to be visited. Hence, a complete information on the system at the <u>user</u> level does not mean a complete information at the <u>model</u> level.

(iii) the time to failure and the repair time are respectively $hyper(\lambda_1, \lambda_2, \alpha)$ and $Erlang(3, \beta, \mu)$ distributed. The corresponding markovian model is shown on Fig. 3.

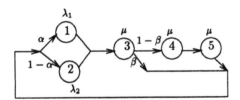

Fig. 3. The 2-hyperexponential failure/3-Erlang repair case.

To describe the system behavior, we have to notice that transitions from state 3 to state 4 and from state 4 to state 5 may occur between a failure instant τ_{2k+1} and a repair instant $\tau_{2(k+1)}$, $k = 0, 1, 2, ...$ Let τ_{2k+1}^1 and τ_{2k+1}^2 be respectively the instants of such transitions. These two instants cannot be observed by the user. Their values are computed as follows. Let T_1, T_2 and T_3 the sojourn times in states 3, 4 and 5 respectively, provided the transitions above occur. We must have $T_1 + T_2 + T_3 = \Delta$, with $\Delta = \tau_{2(k+1)} - \tau_{2k+1}$. It yields

$$\hat{T}_i = E[T_i / T_1 + T_2 + T_3 = \Delta] = \Delta/3 \quad , i = 1, 3.$$

To see this, observe that the $T_i's$ are independent and exponentially distributed with the same parameter μ and that $\hat{T}_1 + \hat{T}_2 + \hat{T}_3 = \Delta$. Therefore, the instants τ_{2k+1}^1 and τ_{2k+1}^2 are estimated by

$$\hat{\tau}_{2k+1}^1 = \tau_{2k+1} + \Delta/3$$

$$\hat{\tau}_{2k+1}^2 = \tau_{2k+1} + 2\Delta/3$$

Observing that $P(\tau_1^1 < \infty) = P(\tau_1^3 < \infty) = 1 - \beta$, the ordered sequence of the jumps of X_t until τ_4 should be :

- $\{\tau_1, \tau_2, \tau_3, \tau_4\}$ with probability β^2
- $\{\tau_1, \tau_1^1, \tau_1^2, \tau_2, \tau_3, \tau_4\}$ or $\{\tau_1, \tau_2, \tau_3, \tau_3^1, \tau_3^2, \tau_4\}$ with probability $(1 - \beta)\beta$
- $\{\tau_1, \tau_1^1, \tau_1^2, \tau_2, \tau_3, \tau_3^1, \tau_3^2, \tau_4\}$ with probability $(1 - \beta)^2$

Hence

$$\mathcal{G}_t = \{\tau_1, X_{\tau_1} = 3, (\tau_1^1 = \infty) \bigcup (\tau_1^1 < \infty)(\tau_1^1, X_{\tau_1^1} = 4, \tau_1^2, X_{\tau_1^2} = 5), \tau_2, X_{\tau_2} \in A_1, \tau_3, X_{\tau_3} = 3, (\tau_3^1 = \infty) \bigcup (\tau_3^1 < \infty)(\tau_3^1, X_{\tau_3^1} = 4, \tau_3^2, X_{\tau_3^2} = 5), \tau_4, X_{\tau_4} \in A_1, \tau_4 \leq t < \tau_5\}$$

The Rule of the Coefficient of Variation

Systems are modeled such as the system state X_t is an homogeneous continuous-time Markov chain. When times to failure and repair times are exponentially distributed, such a modelling is obvious. When they have a phase-type distribution, the markovian representations of such a distribution (cf Fig. 3. for example) lead to a similar result. Otherwise, three steps are performed :

-step 1: identify distributions of the quantities above. A rule based on the coefficient of variation is applied as follows. Let Y be a positive random variable having a finite mean $E[Y]$ and a finite standard deviation σ_Y. Let

$$C_Y = \frac{\sigma_Y}{E[Y]}$$

be the coefficient of variation. Decisions are (cf [7]) :

- if $\frac{1}{k} \leq C_Y^2 < \frac{1}{k-1}$, use the generalized k-Erlang distribution $Erlang(k, \beta, \mu)$,
- if $C_Y^2 = 1$, use the exponential distribution,
- if $C_Y^2 > 1$, use the 2-hyperexponential distribution $hyper(\lambda_1, \lambda_2, \alpha)$.

-step 2: estimate the parameters of the chosen distribution.

-step 3: build a markovian model using when necessary the markovian representations as pointed on Fig. 2 and on Fig. 3.

Reliability Measures in the Operational Phase

Let $\Gamma \subset E$ be the set of operational states of the system state X_t, i.e. the states for which the system is considered operational. For safety systems, inside the set of the non-operational states $E - \Gamma$ (for which the system is halted), we distinguish two subsets:

- the set of benign states Υ, $\Upsilon \subset E - \Gamma$ corresponding to a benign interruption for which system restoration is performed,
- the set of catastrophic states corresponding to a catastrophic interruption, where no system restoration is done. This results from the fact that the consequences of such an interruption are such that system restoration is much less important than the repair of the consequences (damage to property, law suits, etc ... and the analysis of the causes (board of inquiry, etc ...) [1].

Given a history $\mathcal{G}_s$ (which results from some history $\mathcal{H}_s$ at the user level as it is indicated above), the reliability measures of the system in the <u>operational phase</u> are :

reliability $R(s,t) = P(X_u \in \Gamma, \ s \le u \le t/\mathcal{G}_s)$
point availability $A(s,t) = \sum_{i \in \Gamma} P(X_t = i/\mathcal{G}_s)$
mean interval availability on $[s,t]$:
$IA(s,t) = \sum_{i \in \Gamma} \frac{1}{t} \int_s^t P(X_u = i/\mathcal{G}_s) du$
steady-state availability $SA(s) = \lim_{t \to \infty} IA(s,t)$
maintenability
$M(s,t) = P(X_t \in \Upsilon, \ X_u \in \Gamma \bigcup \Upsilon, \ s \le u \le t/\mathcal{G}_s)$
safety $S(s,t) = P(X_u \in \Gamma \bigcup \Upsilon, \ s \le u \le t/\mathcal{G}_s)$

When $s = 0$, quantities above are the usual reliability measures in the <u>design phase</u>. Expressions above have the form $E[f(X_t)/\mathcal{G}_s]$, $s < t$, for some function f depending on the wanted transient characteristic. Assume that X_t is a continuous-time Markov chain. Using the Markov property yields

$$E[f(X_t)/\mathcal{G}_s] = \sum_{j \in E} E[f(X_t)1(X_s = j)/\mathcal{G}_s]$$

$$= \sum_{j \in E} E[f(X_t)/X_s = j]P(X_s = j/\mathcal{G}_s)$$

Suppose now that X_t is time-homogeneous. We have $E[f(X_t)/X_s = j] = E[f(X_{t-s})/X_0 = j]$. Such a term depends only on the initial state and can be computed with classical methods. So that what we have to really compute using past behavior is $P(X_s = j/\mathcal{G}_s)$. For example, the availability in operational phase $A(s,t)$, for which $f(i) = 1(i \in \Gamma)$, can be computed as

$$A(s,t) = \sum_{j \in E} P(X_{t-s} \in \Gamma/X_0 = j)P(X_s = j/\mathcal{G}_s)$$

$$= \sum_{i \in \Gamma} \sum_{j \in E} P(X_{t-s} = i/X_0 = j)P(X_s = i/\mathcal{G}_s)$$

We give below the mathematical background needed to compute reliability measures once the markovian model is build.

BACKGROUND MATERIAL

Point Processes Associated with a Markov Chain

We first recall in this chapter some results of the point processes theory. We refer to Bremaud ([2]) for a complete presentation. Let N_t be a point process adapted to some history $(\mathcal{F}_t)$. N_t admits the $(P, \mathcal{F}_t)$-intensity λ_t if
(i) λ_t is a non-negative $(\mathcal{F}_t)$-progressive process such that for all $t \ge 0$

$$\int_0^t \lambda_s ds < \infty \quad P - a.s$$

(ii) for all non-negative $(\mathcal{F}_t)$-predictable processes C_t,

$$E\left[\int_0^\infty C_s dN_s\right] = E\left[\int_0^\infty C_s \lambda_s ds\right]$$

Let now X_t be a standard homogeneous Markov chain taking its values in an enumerable space E, stable and conservative, i.e.

$$for \ all \ i \in E, \quad q_i = \sum_{i \ne j} q_{ij} < \infty$$

where $(q_{ij}; i, j \in E)$ is the infinitesimal generator of X_t. Let $(\mathcal{F}_t)$ be some history of X_t and $N_t(i,j)$ be the number of transitions from i to j in the interval $(0,t]$. From Levy's formula, we conclude that $N_t(i,j)$ has the $(P, \mathcal{F}_t)$- intensity $q_{ij}1(X_t = i)$. Define

$$\widetilde{E \times E} = \{(i,j); i, j \in E/q_{ij} > 0\}$$

and let θ be a subset of $\widetilde{E \times E}$. We denote $\theta(.,i) = \{j, j \in E/ \ (j,i) \in \theta\}$ and $\theta(i,.) = \{j, j \in E/ \ (i,j) \in \theta\}$. The flow inside θ or the number of transitions in θ during $(0, t]$ is

$$N_t = \sum_{j \in E, i \in \theta(.,j)} N_t(i,j)$$

who admits the $(P, \mathcal{F}_t)$-intensity
$\sum_{j \in E, i \in \theta(.,j)} q_{ij}1(X_t = i)$.

Integral Representation of a Markov Chain

By convention, for a stochastic process Y_t, $\Delta Y_t = Y_t - Y_{t-}$ where Y_{t-} is the left-hand limit of Y at t. Define $Z_t(i) = 1(X_t = i)$ for $i \in E$. It yields

$$Z_t(i) = Z_o(i) + \sum_{o < s \le t} \Delta Z_s(i)$$

where $\Delta Z_s(i) = \sum_{j \in E}(\Delta N_s(j,i) - \Delta N_s(i,j))$.
Let $\bar{N}_t(i,j)$ be the Poisson process with a $(P, \mathcal{F}_t)$-intensity q_{ij}, $i, j \in \widetilde{E \times E}$. We notice that

$$N_t(i,j) = \int_o^t Z_s(i) d\bar{N}_s(i,j)$$

Assume that the $\bar{N}_t(i,j)$'s are without common jumps and that X_o and the processes $\bar{N}$ are mutually independent. Then we derive the following integral representation $Z_t(i) = Z_o(i) + \sum_{j \in E} \int_o^t Z_s(j) d\bar{N}_s(j,i)$

$$- \sum_{j \in E} \int_o^t Z_s(i) d\bar{N}_s(i,j) \qquad (2)$$

Now, let $(\theta_p, \ 1 \le p \le m)$ be a partition of $\widetilde{E \times E}$. Therefore, (2) writes equivalently as

$$Z_t(i) = Z_o(i) + \sum_{1 \le p \le m} \sum_{j \in \theta_p(.,i)} \int_o^t Z_s(j) d\bar{N}_s(j,i)$$

$$- \sum_{1 \le p \le m} \sum_{j \in \theta_p(i,.)} \int_o^t Z_s(i) d\bar{N}_s(i,j) \qquad (3)$$

Filtering Formulas

Explicit filtering formulas for the transient behavior of Markov chains using past histories are given in this section. They can be derived from the filtering equations in [2], [6]. However simple proofs are given here for the readers who are not familiar with filtering theory. Let $\mathcal{G}_t$ a history in terms of X_t. We denote by $P_{\mathcal{G}_t}$ the restriction of P on the probability space $(\Omega, \ \mathcal{G}_t)$. Let τ_n be the last jump of X_t before the time t. We want to compute $P(X_t = i/\mathcal{G}_t)$ on $\{\tau_n \le t < \tau_{n+1}\}$ or equivalently

$$P(X_t = i/\mathcal{G}_t, \ \tau_n \le t < \tau_{n+1})$$

which is equal to

$$\frac{P_{\mathcal{G}_t}(X_t = i, \ \tau_n \le t < \tau_{n+1})}{P_{\mathcal{G}_t}(\tau_n \le t < \tau_{n+1})}$$

We shall consider two kinds of histories: the history where only the jump times of the chain X_t are observed and the history where moreover specific blocks of states are observed. By convention, we write A, B for $A \bigcap B$.

Only the jump times are observed. Let $(\tau_i)_{i \geq 1}$ be the ordered sequence of the jumps of X_t. Here,

$$\mathcal{G}_t = \sigma(\tau_i, \ \tau_i \leq t) \tag{4}$$

is the σ-algebra generated by the random variables $(\tau_i, \ \tau_i \leq t)$. Let $\underline{i}_n = (i_0, i_1, \ldots, i_n)$ be a path of size $n + 1$ of the chain X_t and $\mathcal{S}^n$ the set of such paths. Define

$$\zeta(\underline{i}_n) = \{X_o = i_0, \Delta \bar{N}_{\tau_1}(i_0, i_1) = 1, \bigcap_{j \in E - i_1} (\Delta \bar{N}_{\tau_1}(i_0, j) = 0),$$

$$\Delta \bar{N}_{\tau_2}(i_1, i_2) = 1, \bigcap_{j \in E - i_2} (\Delta \bar{N}_{\tau_2}(i_1, j) = 0), ..., \Delta \bar{N}_{\tau_n}(i_{n-1}, i)$$

$$= 1, \bigcap_{j \in E - i} (\Delta \bar{N}_{\tau_n}(i_{n-1}, j) = 0), \bigcap_{j \in E - i} (\Delta \bar{N}_{t - \tau_n}(i, j) = 0)\}$$

Due to the integral representation (2), it yields

$$\{X_t = i, \ \tau_n \leq t < \tau_{n+1}\} = \bigcup_{\underline{i}_n \in \mathcal{S}^n} \zeta(\underline{i}_n)$$

Let

$$a_k = (\tau_{k+1} - \tau_k) q_{i_k i_{k+1}} e^{-(\tau_{k+1} - \tau_k) q_{i_k i_{k+1}}}$$

for $k < n$ (with $\tau_o = 0$) and $a_n = e^{-(t - \tau_n) q_{i_n}}$.
Since X_o and the processes $\bar{N}$ are mutually independent, it follows that

$$P_{\mathcal{G}_t}(X_t = i, \ \tau_n \leq t < \tau_{n+1}) = \sum_{\underline{i}_n \in \mathcal{S}^n, \ i_n = i} P(i_0) \prod_{0 \leq k <= n} a_k$$

We compute similarly $P_{\mathcal{G}_t}(\tau_n \leq t < \tau_{n+1})$ as

$$P_{\mathcal{G}_t}(\tau_n \leq t < \tau_{n+1}) = \sum_{j \in E} P_{\mathcal{G}_t}(X_t = j, \ \tau_n \leq t < \tau_{n+1})$$

Let

$$L_t(i) = \sum_{\underline{i}_n \in \mathcal{S}^n, \ i_n = i} P(i_0) \prod_{0 \leq k <= n} b_k$$

with $b_k = a_k / ((\tau_{k+1} - \tau_k))$. Hence

$$P(X_t = i / \mathcal{G}_t) = \frac{L_t(i)}{\sum_{j \in E} L_t(j)} \tag{5}$$

on $\{\tau_n \leq t < \tau_{n+1}\}$.

The jump times and blocks of states are observed. Let $(\theta_p, \ 1 \leq p \leq m)$ be m mutually disjointed subsets of $\widetilde{E \times E}$ such that $\bigoplus_{1 \leq p \leq m} \theta_p = \widetilde{E \times E}$, where by convention $A \bigoplus B$ holds for $A \bigcup B$ and $A \bigcap B = \emptyset$. We denote by $N_t^{(p)}$ the traffic flow in θ_p and by $\hat{\lambda}_t^{(p)}$ the $(P, \mathcal{F}_t)$-intensity of $N_t^{(p)}$, $1 \leq p \leq m$. Let

$$\mathcal{G}_t = \sigma(N_s^{(p)}, 1 \leq p \leq m, \ 0 \leq s \leq t) \tag{6}$$

be the common history of all the $N_t^{(p)}$'s. Define a sequence of random variables $(v_n, \ n \geq 1)$ on $(\Omega, \ \mathcal{F}, \ P)$, with values in $\{1, m\}$ and such that for all n

$$v_n = p, \quad if \ \Delta N_{\tau_n}^{(p)} = 1 \tag{7}$$

We notice that the $N_t^{(p)}$ are without common jumps, the family $(\theta_p, \ 1 \leq p \leq m)$ being a partition.
Let $\underline{i}_n^{\theta, \mathbf{Y}_n} = (i_0, i_1, \ldots, i_n)$ be a path of size $n + 1$ of the chain X_t such as $i_k \in \theta_{v_k}(., i_{k+1}), 0 \leq k < n$ and $\mathcal{S}_n^{\theta, \mathbf{Y}_n}$ the set of such paths.
From (3), it yields

$$\{X_t = i, \ \tau_n \leq t < \tau_{n+1}\} = \bigcup_{\underline{i}_n \in \mathcal{S}_n^{\theta, \mathbf{Y}_n}} \zeta(\underline{i}_n)$$

Let

$$L_t^{\theta, \mathbf{Y}_n}(i) = \sum_{\underline{i}_n \in \mathcal{S}_n^{\theta, \mathbf{Y}_n}, \ i_n = i} P(i_0) \prod_{0 \leq k <= n} b_k$$

with $b_k = a_k / ((\tau_{k+1} - \tau_k))$. Therefore, a similar proof gives

$$P(X_t = i / \mathcal{G}_t) = \frac{L_t^{\theta, \mathbf{Y}_n}(i)}{\sum_{j \in E} L_t^{\theta, \mathbf{Y}_n}(j)} \tag{8}$$

on $\{\tau_n \leq t < \tau_{n+1}\}$.

Remark

The formula (8) above holds when the values of the random variables $(v_i, \ 1 \leq i \leq n)$ defined in (7) are perfectly known. It means that at each instant τ_i, we know the unique p $(p \in \{1, 2, \ldots, m\})$ for which $\Delta N_{\tau_i}^{(p)} = 1$, $1 \leq i \leq n$. Assume now that such an information is incomplete, i.e. at the instant τ_i, $\Delta N_{\tau_i}^{(p)} = 1$ for some (unique) $p \in V_i$ where V_i is a subset of $\{1, 2, \ldots, m\}$, $1 \leq i \leq n$. Then formula (8) writes

$$P(X_t = i / \mathcal{G}_t) = \frac{\sum_{v_k \in V_k, 0 \leq k < n} L_t^{\theta, \mathbf{Y}_n}(i)}{\sum_{j \in E} \sum_{v_k \in V_k, 0 \leq k < n} L_t^{\theta, \mathbf{Y}_n}(j)} \tag{9}$$

on $\{\tau_n \leq t < \tau_{n+1}\}$.

In what follows, we present and show use of the reliability tool SOLAR.

THE RELIABILITY TOOL SOLAR

Features and Design of SOLAR

SOLAR (Systems On Line Analyser for Reliability) is written in C language and runs on the UNIX 4.3BSD system, on SUN stations. This should minimize machine transportation difficulties. The software product provides four modules : input module, edit module, reliability menu module, plotting routines module.

1. The input module supports entering data options, from a file or from the terminal emulated in an edit window.

2. The edit module is devoted to data manipulations, like insert, change, delete and list data. It is represented by the text subwindow package of Sunview Tools.

3. The reliability menu allows various reliability measures to be computed with or without using past behavior.

4. The plotting module supports displaying reliability data versus time t.

<u>The Syntaxic Language of SOLAR</u>

Such a tool needs two input files written in a syntaxic language similar to the one used in [3], [4]. The first input file, denoted by DESIGNF, describes the system in terms of the nature of components, the known parameters and their value, the component failure and repair rates, the dependencies between components, the minimum numbers of components under which the system is considered operational (respectively safe) and the repair strategy.

The second input file, denoted by DATAF, includes the history of the system up to the final observation time. It consists on the available components and their number at the initial observation time, the past events such as failures or repairs and times at which they occurs and the value of the final observation time. The syntax of the language is shown through the examples below.

Example 1

Let us consider a multiprocessor with two processors and three memory units connected to a bus, as indicated on Fig. 4. A failure of a component is assumed to be independant of failures of any other component in the system. Each processor may fail in two different modes, "soft" fail and "hard" fail. If a soft fail occurs, the failed processor has to be restarded to resume operation. The system is considered up if at least one processor, two memory units and the bus are operational.

The following repair strategy is adopted. If the system goes down due to a hard fail, the component which brought it down has (preemptive) priority of being repaired. If the system remains operational after a hard fail (necessarily not the bus), the processor has (preemptive) priority over the memory. It is assumed that no failures can occur when the system is down.

For this system, failure rates are 1 per 10 days (240 hours) for a memory unit, 1 per 3 months for the bus and 1 per 5 days (120 hours) for a processor. 90% of processor failures come from software. We assume that there is only one repairman available for hard failures. It takes an average of 8 hours to repair the hardware part of a processor and an average of 4 hours to repair a memory unit or a bus. To restart a processor for which the software fails takes an average of 1 hour and 15 minutes.

In the **DESIGNF** file, the user enters :
- **the component types and their number**
SYSTEM
mem 3
proc 2
soft 2
bus 1

- **the known parameters and their value**

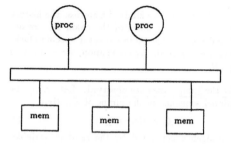

Fig. 4. The bus, memory, processor example

PARAMETERS
memfr 0.00417
memrr 0.25
....

The parameters are usually the quantities which describe the failure or repair rates of the system units. If a parameter, say the memory repair rate memrr, is unknown, we write
memrr unknown

- **the component characteristics such as failure and repair rates.**
COMPONENT : mem
FAILURE RATE : memfr
REPAIR RATE : memrr

which holds for the memory component with an exponential failure (repair) rate with parameter memfr (memrr). Let now the memory component have a generalized k-Erlang failure holding time with parameters β and memfr and a 2-hyperexponential repair holding time with parameters memrr1, memrr2 and α. Then the syntax is
COMPONENT : mem
FAILURE RATE : Erlang(k,β,memfr)
REPAIR RATE : hyper(memrr1,memrr2,α)

If a distribution, say the repair distribution, is unknown, we write
REPAIR RATE : unknown
SOLAR estimates unkown parameters or unknown distributions, given the past history of the system.

- **the dependencies between components**
The general formulation writes using a combination of the components and the two operators 'and', 'or'. For example : (unit i or unit j) and (unit k or unit l).... Here, we have
DEPENDENCIES
proc and soft

- **the minimum numbers of components under which the system is considered operational**
RELIABILITY EVALUATION
mem 2
proc 1
soft 1
bus 1

- **the minimum numbers of components under which the system is considered safe**
SAFETY EVALUATION

- **the repair strategy**, i.e. the components which have to be repaired with priority.

REPAIR STRATEGY

mem 1

proc 2

soft 2

bus 1

The number 2 runs for the highest priority. In the **DATAF** file, the user enters
- **the available component types and their number at the beginning time of observation**

CURRENT SYSTEM

mem 3

proc 2

soft 2

bus 1

- **the instant at which the user begins to observe the system**

INITIAL OBSERVATION TIME 0.00

- **the past events** such as failures or repairs and the instants at which they occurs

EVENTS

failure(mem) 0.74

repair(mem) 10.23

failure 27.09

failure 27.42

- **the final observation time**

FINAL OBSERVATION TIME

29.00

The actions of histories on reliability and availability are shown on Table 1.

with past actions			with no past actions		
time	relia-bility	availa-bility	time	relia-bility	availa-bility
29.39	0.694	0.738	4.69	0.999	0.997
30.56	0.694	0.824	6.25	0.999	0.996
31.34	0.694	0.858	9.38	0.999	0.995
33.69	0.694	0.916	12.50	0.999	0.995
41.50	0.694	0.974	18.75	0.999	0.993
47.75	0.694	0.983	25.00	0.999	0.993
54.00	0.694	0.986	37.50	0.999	0.991
66.50	0.694	0.987	50.00	0.999	0.990
79.00	0.694	0.987	75.00	0.999	0.989
104.00	0.694	0.987	100.00	0.999	0.988

Table 1. System reliability and availability.

Example 2

Assume now that the multiprocessor above is the safety part of a dynamic system as indicated on Fig. 5. The crash of the multiprocessor is viewed here as a catastrophic failure, i.e. no restoration is done when the multiprocessor is down.

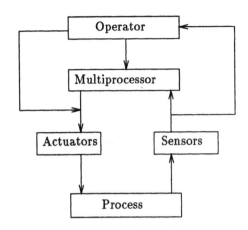

Fig. 5. A safety system.

In the **DESIGNF** file, for the parts concerning the component types with their number and the reliability evaluation, the user may enter:

SYSTEM

mem 3

proc 2

soft 2

bus 1

operator 1

actuator 2

process 1

sensor 2

....

RELIABILITY EVALUATION

mem 2

proc 1

soft 1

bus 1

operator 1

actuator 1

process 1

sensor 2

But for the parts concerning the safety evaluation, since safety concerns only the multiprocessor the user enters:

SAFETY EVALUATION

mem 2

proc 1

soft 1

bus 1

The actions of histories on maintenability and safety are shown on Table 2.

The Underlying Markov Chains

The corresponding Markov chains are automatically constructed by SOLAR. For the chain corresponding to the first example, a state is first represented by an 4-uplet $n_1 n_2 n_3 n_4$ where n_1 is the number of failed memories, n_2 the number of failed processors, n_3 the number of failed softwares and n_4 the number of failed bus.

with past actions			with no past actions		
time	mainte-nability	safety	time	mainte-nability	safety
29.39	0.436	0.999	4.69	0.341	0.999
30.56	0.274	0.999	6.25	0.334	0.999
31.34	0.257	0.999	9.38	0.322	0.999
33.69	0.258	0.999	12.50	0.313	0.999
41.50	0.275	0.999	18.75	0.304	0.999
47.75	0.281	0.999	25.00	0.299	0.999
54.00	0.286	0.999	37.50	0.296	0.999
66.50	0.296	0.999	50.00	0.295	0.999
79.00	0.307	0.999	75.00	0.297	0.999
104.00	0.330	0.999	100.00	0.303	0.999

Table 2. System maintenability and safety.

The state space is
$S = \{0000, 0010, 0020, 0100, 0110, 0200, 1000, 1010,$
$1020, 1100, 1110, 1200, 2000, 2010, 2100, 1101, 1011,$
$1001, 0101, 0011, 0001\}.$

For conveniance, these states are numbered respectively $0, 1, ..., 20$, i.e. $E = \{0, 1, .., 20\}$ is prefered to be the state space of X_t instead of S.

In what follows, we first translate the history above which writes

$\mathcal{H}_t = \{0.74, \; failure(mem), \; 10.23, \; repair(mem),$
$27.09, \; failure, \; 27.42, \; failure\}$

in terms of X_t. Second, we point out which formula has to be applied to compute the reliability and availability measures. In this history, a second failure is detected before the isolation of the previous failed unit. Hence, the steps before the repair step take a non-negligible time.

We construct the θ_p's used in (8) as follows. Let $\gamma_1, \gamma_2, \gamma_3, \gamma_4$ be the subsets of couples of states corresponding respectively to the failure of a memory, a processor, a soft and a bus (cf appendix A) :

$\gamma_1 = \{(0, 6), (1, 7), (3, 9), (6, 12), (7, 13), (9, 14)\}$
$\gamma_2 = \{(0, 3), (1, 4), (3, 5), (6, 9), (7, 10), (9, 11)\}$
$\gamma_3 = \{(0, 1), (1, 2), (3, 4), (6, 7), (7, 8), (9, 10)\}$
$\gamma_4 = \{(0, 20), (1, 19), (3, 18), (6, 17), (7, 16), (9, 15)\}$

The subsets $\delta_1, \delta_2, \delta_3, \delta_4$ of couples of states corresponding to the repair of a memory, a processor, a soft and a bus are respectively (cf appendix A) :

$\delta_1 = \{(6, 0), (12, 6), (13, 7), (14, 9)\}$
$\delta_2 = \{(3, 0), (4, 1), (5, 3), (9, 6), (10, 7), (11, 9)\}$
$\delta_3 = \{(1, 0), (2, 1), (4, 3), (7, 6), (8, 7), (10, 9)\}$
$\delta_4 = \{(15, 9), (16, 7), (17, 6), (18, 3), (19, 1), (20, 0)\}$

We define the θ_p's by $\theta_1 = \gamma_1, \; \theta_2 = \gamma_2, \; \theta_3 = \gamma_3,$
$\theta_4 = \gamma_4, \; \theta_5 = \delta_1, \; \theta_6 = \delta_2, \; \theta_7 = \delta_3, \; \theta_8 = \delta_4.$
Therefore, the history above writes (using the same notation) at the model level:

$\mathcal{G}_t = \{0.74, \; X_{0.74} \in \theta_1, \; 10.23, \; X_{10.23} \in \theta_5, \; 27.09,$

$$X_{27.09} \in \bigoplus_{1 \leq p \leq 4} \theta_p, \; 27.42, \; X_{27.42} \in \bigoplus_{1 \leq p \leq 4} \theta_p\}$$

Reliability measures in the operational phase are then computed as pointed out in the first chapter using the formula (9), with $V_1 = \{1\}, \; V_2 = \{5\}, \; V_3 = \{1, 2, 3, 4\}, \; V_4 = \{1, 2, 3, 4\}.$

A similar analysis can be applied to the chain corresponding to the second example above.

Remark

A choice is let to the user to enter the characteticts of the system and the history in terms of states of the Markov chain. Instead of the DESIGNF and DATAF files, the user enters the DESIGNF1 and DATAF1 files given in appendix A. Due to the Markov property, the history is entered only from the most recent observed value (when it exists) of X_t, which here is $X_{10.23} = 0$.

CONCLUSIONS

A software tool dedicated to the quantitative evaluation of dependability measures of fault-tolerant systems is presented. The problem addressed concerns the actions of past behavior of such systems during their operational phase. The most accurate and updated predictions are available to the user at the time of implementation on the control on the basis of collected data. This allows the user to take optimal decisions for a reliable functioning of the system.

REFERENCES

[1] J. Arlat, K. Kanoun. Modelling and dependability evaluation of safety systems in control and monitoring applications. Safecomp '86, Sarlat-France, pp 157-164.

[2] Bremaud, P. Point processes and queues, Martingales dynamics. Springer-Verlag, New york Inc., 1981.

[3] A. Goyal, S.S. Lavenberg. Modelling and analysis of computer system availability. RC 12458 T.J. Watson research center, Yorktown Hts, NY 1986.

[4] A. Goyal. System availability estimator (SAVE), User's manual. RC 12517 T. J. Watson research center, Yorktown Hts, NY 1987.

[5] D.A. Siewiorek. Fault tolerance in commercial computers. IEEE review Computer, July 90, pp 26-37

[6] K. Sismail. SOLAR : Un outil de prédiction en fiabilité pour les systèmes on ligne. Proc.7th Int. Conf. on reliab. and maintenability, Brest-France, 18-22 June 1990, pp 96-104.

[7] H. C. Tijms. Stochastic modelling and analysis: a computational approach John Wiley and Sons, 1986, pp 393-400.

A The Multiprocessor Markov Chain Representation

1. The DESIGNF1 file :
Number of states
21
Operational states
0 1 3 6 7 9
Catastrophic states
none
Parameters

memfr 0.00417
memrr 0.25
procfr 0.0.000833
procrr 0.125
softfr 0.0075
softrr 0.800000
busfr 0.000463
busrr 0.25

Matrix

0–1 2*softfr	0–3 2*procfr	0–6 3*memfr
0–20 busfr	1–0 softrr	1–2 softfr
1–4 2*procfr	1–7 3*memfr	1–19 busfr
2–1 softrr	3–0 procrr	3–4 2*softfr
3–5 procfr	3–9 3*memfr	3–18 busfr
4–1 procrr	4–3 softrr	5–3 procrr
6–0 memrr	6–7 2*softfr	6–9 2*procfr
6–12 2*memfr	6–17 busfr	7–6 softrr
7–8 softfr	7–10 2*procfr	7–13 2*memfr
7–16 busfr	8–7 softrr	9–6 procrr
9–10 2*softfr	9–11 procfr	9–14 2*memfr
9–15 busfr	10–7 procrr	10–9 softrr
11–9 procrr	12–6 memrr	13–7 memrr
14–9 memrr	15–9 busrr	16–7 busrr
17–6 busrr	18–3 busrr	19–1 busrr
20–0 busrr		

Initial states and their probability
0 1.00

2. The DATAF1 file:
Initial time
10.23
Initial states and their probability
0 1.
Number of observations
2
Block sequences, their cardinal and their states
no=1 cardinal=4 states= 1 3 6 20
no=2 cardinal=9 states= 2 4 7 19 5 9 18 12 17
Occurrence times and concerned blocks
27.09 1
27.42 2
Final observation time   29.00

APPLYING PROGRAMMABLE GATE ARRAYS TO PROVIDE DIVERSITY IN SAFETY RELEVANT SYSTEMS

W. A. Halang and J. M. Schut

*Dept. of Computing Science, University of Groningen,
P.O. Box 800, 9700 AV Groningen, The Netherlands*

Abstract. Although there are already a number of established methods and guidelines, which have proven their usefulness for the development and verification of high integrity software employed for the control of safety related technical processes, these measures cannot ultimately guarantee the correctness of larger programs with mathematical rigour, yet, and, hence, corresponding safety licences are generally denied. As a remedy for this unsatisfactory situation, a novel approach based on the utilisation of programmable gate arrays is presented. The hard-wired discrete logic built of relays or LSI/MSI-chips, which operates in many safety related control systems in parallel to computers, is replaced by programmable logic constructed from PGAs. Thus, the flexibility of programmable electronic systems is combined with the long established and generally accepted rigorous certifyability of hard-wired sequential circuits, since in both cases the same design and verification procedures are employed. A PGA is programmed to perform certain Boolean or sequential functions by loading appropriate bit patterns into internal static memory cells. A method for the verification of their contents is detailed, which is based on diverse backward analysis and which can easily be automated. The utilisation of PGAs is not only advantageous for replacing hard-wired by programmable logic, but also as a novel means of providing diverse redundancy in programmable electronic systems comprising computers and programmable logic controllers.

Keywords. Programmable logic; programmable electronic systems; safety critical automation; high integrity systems; emergency shutdown systems; safety licensing; diverse backward analysis; diverse redundancy.

INTRODUCTION

Highly safety critical control systems, such as emergency shutdown systems, are still a stronghold of hard-wired logic. This is due to the fact that the licensing authorities can rely on procedures for their verification, which are well-established for already a long time. The main disadvantage of hard-wired systems is their inflexibility. In industry, the advent of flexible manufacturing systems brought about an urgent need that the safety systems should be adjustable to varying requirements with the same ease as the proper production equipment.

In order to provide the flexibility needed, it is an obvious possibility to put programmable electronic systems, i.e. computers and programmable logic controllers, in charge of safety functions. The latter are then carried out by software, which can be modified with great ease. There are already a number of established methods and guidelines, which have proven their usefulness for the development of high integrity software employed for the control of safety related technical processes (Bishop, 1990; Clutterbuck and Carré, 1988; DGQ-NTG, 1986; EWICS, 1985; Faller, 1988; Grimm, 1985; Hölscher and Rader, 1984; IEC, 1986; Krebs, 1984; Krebs and Haspel, 1984; O'Neill and others, 1988; Redmill, 1988, 1989; Schmidt, 1988). Prior to its application, such a software is further subjected to appropriate measures for its verification and validation (loc. cit.). However, according to the present state of the art, these measures cannot ultimately guarantee the correctness of larger programs with mathematical rigour. For this reason, the safety licensing authorities are very reluctant in approving highly safety critical systems, whose behaviour is exclusively software controlled. If the complexity of such control software is non-trivial, the safety licence is generally denied, yet. As a result, it is a common feature of industrial practice to find programmable electronic systems controlling technical processes, and working in parallel to hard-wired logic in charge of the safety functions, such as emergency shutdown. In case of a conflict, the hard-wired logic always overrides the computer's actions.

To provide a remedy for this unsatisfactory situation, it is the purpose of this paper to present a novel approach to solve the problem outlined above, which became possible by employing the new technology of programmable gate arrays (PGAs; synonymously called logic cell arrays, LCAs, as well), which was launched by Xilinx Inc. in 1985 with the XC2000 series. The hard-wired relay or LSI/MSI-chip based discrete logic in charge of safety related control functions is to be replaced by programmable logic constructed from PGAs. A PGA is programmed to perform certain Boolean or sequential functions by loading appropriate bit patterns into internal static memory cells. Thus, the flexibility of a programmable electronic device is combined with the long established and generally accepted rigorous certifyability of hard-wired sequential circuits, since in both cases the same design and verification procedures are employed. It is a psychological advantage, whose value should not be underestimated, that our approach does not require from the licensors to step directly from one paradigm of system implementation to a totally different one, viz. from combinational gates and single bit storage elements working in parallel to sequential program code. Instead, a smooth transition is provided by using a programmable technology and, at the same time, adhering to the design and verification procedures to which the licensors are well accustomed.

Thus, PGAs are suitable to replace hard-wired logic in safety critical control systems, also in those environments where they work in parallel with computers. We expect that they will persist in this rôle, even after rigorous methods for software licensing will have been widely established, namely as a means of providing diverse redundancy in programmable electronic systems.

This paper is organised as follows. We begin with a short description of PGAs, their function, and their capabilities. Then, we give an outline of how PGAs are programmed and how this is facilitated by CAD tools. For our purposes a subset of the functionality provided by these tools is identified, in order to enable straightforward safety licensing of PGA programs. Thus, placing and routing optimalisation is renounced and composite functional modules, which are taken from a library, are always inserted by hand following a priori verified design schemes. The mentioned restrictions ensure that the bit patterns generated to let PGAs perform specified functionalities become easily predictable. These bit patterns are stored in non-volatile read only memories. The final section of this paper deals with the most important aspect of our approach, viz. the rigorous verification of a given function's implementation. This is achieved in two steps. First, the correctness of a corresponding combinational or sequential circuit is established with the traditional procedures. Secondly, the mapping of such a circuit on the programmable internal function blocks of a PGA is verified by diverse backward analysis of the pertaining bit pattern read out of the earlier produced ROM. This analysis can be greatly facilitated by automated tools.

PROGRAMMABLE GATE ARRAYS

Programmable gate arrays are modern microelectronic devices, which can be configured by the users to perform all Boolean and sequential functions. A PGA is able to substitute for a printed circuit board with a large number of MSI/LSI-chips. The particular functionality of a PGA is determined by a 'Configuration Program' (CP), which is loaded into on-chip static memory cells of the PGA upon applying power, and which may be replaced at any time to implement a different function. Digital logic is designed and tested with the help of special software. These CAD tools also transform the resulting circuitry diagrams into configuration programs for PGAs. The programs can either be stored in memories with serial or byte-parallel organisation or they can be down-loaded from a host computer. Such a program is not an instruction sequence as known from computers, but a bit pattern defining Boolean functions in the form of truth tables, selecting options, and switching internal communication links. So, it is possible for the PGA to implement higher functions, such as adders, counters, multiplexers etc.

The first generations of programmable logic devices were Programmable Array Logic (PAL) and Generic Array Logic (GAL). They consist of simple configurable logic. Internally, a PAL is divided into an And-matrix, an array of And-gates, an Or-matrix, and an array of Or-gates. The logic functions And and Or are static ones, i.e. they cannot be changed. The function of a PAL-chip can be programmed once by fusing a number of joints in the And- or the Or-matrix, depending on the type of the chip. The general internal structure of a GAL only differs slightly from the one of a PAL-device. The logic function of GALs, however, is more complex. The above-mentioned arrays of And- and Or-gates are replaced by one array of logic function blocks, called Output Logic Macro Cells (OLMCs), which can be configured by the user.

As one of the latest developments in programmable logic, programmable gate arrays are much more complex than PALs and GALs and, therefore, are able to implement more numerous or more complex functions on one chip. They do not allow for more complex (programmable) Boolean functions than GALs, but it is possible to interconnect the various logic function blocks in a much more sophisticated way imposing almost no restrictions.

In the following section we shall describe the internal structure of PGA devices.

INTERNAL STRUCTURE OF PGAs

Programmable gate arrays are provided by different manufacturers in a variety of designs, types, environmental specifications, pinnings, and packagings. The PGA typical parameters, however, are the number of their Configurable Logic Blocks (CLB) and their

Input/Output Blocks (IOB), and the complexity of the CLBs, the IOBs, and of the internal connectivity called Programmable Interconnect (PI). The functionality of these structural components can be configured by means of the above mentioned program, which is held internally in the Configuration Memory (CM).

A schematic representation of a CLB's internal structure is given in Fig. 1. The main element of a CLB is a user configurable Combinatoric Function (CF). Such a CF operates on a number of input variables, some of which may be internal ones, i.e. lines fed back from the CF's output. The outputs generated by a CF can be stored in D-flip-flops and passed on to one of the (internal) output connectors of the CLB. For example, in the Xilinx XC3000 series of PGAs (Xilinx, 1989), a CLB consists of a 4- or 5-input CF, which is implemented by a truth table, and 2 D-flip-flops.

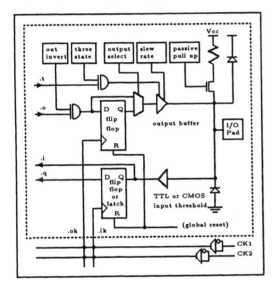

Fig. 1. Internal structure of a CLB.

Fig. 2 displays the internal structure of an IOB. Each IOB is connected to one of the chip's pins. By a number of bits in the configuration program each IOB is individually configurable as an input or as an output port. In the input mode the IOB works as a tri-state-buffer, and the signal present at the corresponding pin can be either buffered or directly passed on to the PI. The output signal can also be buffered and optionally provided in inverted form.

The programmable interconnect represents a system of connections between CLBs and IOBs and between two CLBs each. There are three types of connections. The first one is called General Purpose Interconnect. It connects blocks through lines which run between the rows and columns of the matrix of CLBs. In the Switching Matrices the connection between two horizontal and/or vertical lines is established. The configuration bits for the switching matrices are held in the configuration program as well. A second possibility is to make use of the so-called Direct Interconnect, which allows to link adjacent blocks with each other.

The third connection type are the Long Lines. These are generally used to distribute global variables, such as reset or clock pulses, to a large number of blocks.

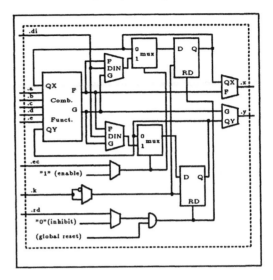

Fig. 2. Internal structure of an IOB.

CONFIGURING PGAs

Configuring PGAs 'by hand' is time-consuming, because already the smallest PGA types contain 64 CLBs and 58 IOBs. Therefore, suitable software tools are provided by the PGA manufacturers supporting and partly automating the design. These CAD tools are based on libraries of predefined standard functions such as counters, comparators, adders, etc. It is possible to incorporate further user-defined components into these libraries. To design a circuitry, which is to perform a given function and which is to be implemented on a PGA, instances of appropriate functional modules are invoked from the library and then interconnected ("wired together"). Thus, there is essentially no difference in the design process as compared to more conventional tools, whose library items correspond to particular MSI/LSI-ICs.

When a circuitry design has been completed, the CAD tool translates it into a net-list file, which describes the components of the circuit and their interconnections. Subsequently, the net-list file is processed by a mapping program. This substitutes the library modules for partial circuits composed of CLBs and, then, associates the latter with CLBs of the PGA. It further places required IOBs and transforms the net-list into linkages via the programmable interconnect. Optionally, the mapping program can optimise the routing of the interconnections, in order to reduce the transmission delays and the power consumption, and to minimise the waste of internal resources.

In the next step the resulting PGA implementation is subjected to a simulator, which allows for design validation and the determination of its timing characteristics.

Finally, the mapped design is converted into a bit-stream which represents the corresponding configuration program. These bit-streams are usually put into (programmable) read only memories or saved by host computers. In the former case, they are read out of the PROM and placed into the PGA's configuration memory upon applying power or upon system initialisation. In the latter case more flexibility is provided: by down-loading a new bit-stream from a host computer to the CM of a PGA its functionality can be modified at any time, i.e. "on the fly"[1].

Besides designing PGA functions with the help of CAD tools, it is also possible to do it 'manually', i.e. to individually define every single bit of the configuration program. This is much more work, but by by-passing the mapping and routing programs the manual approach, which can be carried out using a Design Editor, provides full transparency of the internal details of a PGA implementation. In contrast to this, the output of the CAD tool does not specify how the library functions should be mapped onto the PGA. It only indicates how many CLBs and IOBs are needed, how these blocks should be configured, and how they are to be interconnected.

HARDWARE PROGRAMMING OF PGAs

When power is applied to a PGA-chip, and its voltage reaches the lower threshold of 3 V, an *internal power-on-reset* circuit is activated. This circuit performs a number of functions, two of them are to clear the configuration memory, and to configure the IOBs in such a way that the I/O pins are put into the high impedance state.

The capacity of the configuration memory is dependent on the PGA type. For the Xilinx XC3020, for example, this capacity is 14,775 bits. Here, a configuration program consists of 197 data frames, each one 75 bits long, and is sent with a preceding length count of 24 bits. When a configuration program is down-loaded to such a PGA, it is enclosed in a preamble and a postamble. The preamble specifies among other the synchronisation bits, which are transmitted between successive data frames.

TABLE 1 Possible Start-Up Modes of a PGA

M0	M1	M2	Clock	Mode	Data
0	0	0	active	master	bit serial
0	0	1	active	master	byte addr. 0000 ↑
0	1	0	—	reserved	—
0	1	1	active	master	byte addr. FFFF ↓
1	0	0	—	reserved	—
1	0	1	passive	periph.	byte wide
1	1	0	—	reserved	—
1	1	1	passive	slave	bit serial

---

[1] This feature can be used to implement a variety of special purpose co-processors with only one hardware module.

The operation of a PGA may be initiated in one of eight different start-up modes, which are selectable with the M0, M1 and M2 pins. In Table 1 an overview is given on the different start-up modes.

There are three major modes in which a XC3020-PGA can start. These and a composite mode are detailed in the following overview.

Master mode There are three different ways for a PGA to load a configuration program in master mode. The first one is via a serial link, the DIN-pin. The data can be sent from a computer to the PGA, or can be loaded from a memory chip with bit-serial organisation, which is a very compact way of storing data. The other alternative is to load the CM byte by byte via the data lines D0 – D7. The transmission process is controlled by the PGA, which places the corresponding addresses on the address lines A0 – A15. The byte addresses may be incremented from 0 or decremented from $FFFF_{16}$ (to make use of unoccupied addresses in (BIOS) PROMs).

Peripheral mode There is only one difference of down-loading in this mode and in the master mode, viz. that the process is not controlled by the PGA but by an external computer. The data are transmitted in byte format.

Slave mode Here data are sent in peripheral mode, but bit-serially.

Daisy chain This fourth form of down-loading does not belong to the main modes. It *may* be used in connection with any of the former modes. After the first of the PGAs forming a daisy chain has received all its program bits (up to the number which was specified in the header of the stream), it sends all further arriving bits via its data-OUT-pin to the next PGA. The transmission is synchronised by a clock signal available at the CCLK-pin.

A loaded configuration program can be read back at all times. This provides the opportunity for its ultimate verification.

ENABLING CERTIFICATION

Regardless of the envisaged implementation of a safety critical control system either by hard-wired logic or by programmable gate arrays, there is no difference in the design and functional verification of such a system. In both cases it is designed as and described in the form of a combinational and/or sequential digital circuit. Certifying a PGA implementation requires to show the correctness of the mapping of such a digital circuit onto the PGA's functional blocks and of their corresponding interconnection. The mapping needs to be carried out manually, since the correctness of the CAD tool's mapping software would otherwise have to be established as well. This, however, is still impossible owing to such a utility's high complexity. The same argument holds for the automatic routing optimisation, too. Furthermore, placing and routing optimisation reduces the understandability of a realisation by increased complexity, thus making safety licensing harder. For the mentioned reasons we renounce automatic mapping and optimisation to facilitate straightforward certifyability at the price of more human work and reduced speed of the PGA implementations. The latter shortcoming, however, is without any significance for the application domain considered, because the speed gains achievable lie in the range of nanoseconds.

As tool supporting his work, the designer can use a library of functional building blocks besides the above mentioned design editor. However, the substitution of a functional block by a predefined pattern of appropriately interconnected CLBs and/or IOBs is not carried out after the design process and invisible to the user as by the automatic tools discussed above, but by direct insertion of the pattern into the circuitry diagram, the designer is working on. It is needless to say, that the library components must have passed an a priori certification. Following the outlined design procedure, it is ensured that the designer has always full control over the assignment of logical functions to hardware devices internal to the PGA. In other words, the restrictions imposed on the design process allow to easily predict the configuration program's bit pattern corresponding to the functionality to be implemented.

VERIFICATION AND CERTIFICATION

Based on the boundary conditions set above, the design and verification process of a safety relevant control system proceeds as follows.

1. A combinational and/or sequential digital circuit realising the control system is designed (on paper) and documented in form of a diagram using standard symbols for composite function blocks.

2. The correctness of the resulting circuit is established by rigorous verification employing the traditional procedures as described in e.g. (Bennetts, 1984; Breuer and Friedman, 1976; Geisselhardt, 1978; Görke, 1973; Reinert, 1979).

3. From the circuitry diagram a PGA implementation is derived with the help of a design editor and a library of certified components, which substitute for the composite function blocks occurring in the diagram. This development step yields documentation exactly specifying the setting of each bit in the configuration memory.

4. The design obtained in the previous step is verified (by human licensors) against the specification, i.e. the diagram produced in the first step.

5. A utility software generates the corresponding configuration program, which is then stored in one or more non-volatile read only memories.

6. The bit-stream contained in these ROMs is read out and compared, by more than one human licensor, with the bit setting obtained in step (3). This analysis is a modification of the diverse backward analysis earlier developed for the verification of computer software (Krebs and Haspel, 1984). It can be greatly facilitated by semi-automatic tools, which are similar to the ones developed to support the backward analysis and presented in (Dahll, Mainka and Märtz, 1988; Anders, Mainka and Rabe, 1990).

In the utilisation phase of the control system the configuration program can either be read into the PGA from a ROM or down-loaded from a host computer. After any loading operation, however, the CP shall be read out of the CM and checked, by a simple comparator, against the contents of a(nother) ROM, before the PGA takes charge of a safety critical process.

CONCLUSION

In this paper, we have described a novel approach for the implementation of control systems in safety relevant applications, which is based on the utilisation of programmable gate arrays. Our approach yields the following advantages:

- it provides the same flexibility as programmable electronic systems,

- the same design methods as for hard-wired control systems are applied,

- these designs are certified with the same procedures as combinational and/or sequential logic,

- the licensors can continue to apply the design and verification paradigms they are familiar with for a long time,

- implementations are verified by the easily conceivable method of diverse backward analysis, and

- it can provide flexible diverse redundancy in programmable control systems.

REFERENCES

Anders, U., E.-U. Mainka, and G. Rabe (1990). Tools and Methodologies for Quality Assurance. In B. K. Daniels (Ed.). *Safety of Computer Control Systems 1990*. IFAC Symposia Series, 1990, No. 17. Pergamon Press, Oxford. pp. 113 – 118.

Bennetts, R. G. (1984). *Design of testable logic circuits*. Addison Wesley, London.

Bishop, P. G. (Ed.) (1990). *Dependability of Critical Computer Systems – 3*. Guidelines produced by The European Workshop on Industrial Computer Systems Technical Committee 7. Elsevier. Amsterdam New York.

Breuer, M. A., and A. D. Friedman (1976). *Diagnosis and reliable design of digital systems.* Computer Science Press, Potomac.

Clutterbuck, D. L., and B. A. Carré (1988). The verification of low-level code. *IEE Software Engineering Journal*, 97 – 111.

Dahll, G., U. Mainka, and J. Märtz (1988). Tools for the Standardised Software Safety Assessment (The SOSAT Project). In W. D. Ehrenberger (Ed.). *Safety of Computer Control Systems 1988.* IFAC Proceedings Series, 1988, No. 16. Pergamon Press, Oxford. pp. 1 – 6.

DGQ-NTG (1986). *Software-Qualitätssicherung.* Schrift 12-51. Beuth-Verlag, Berlin.

EWICS TC7 Software Sub-group (1985). Techniques for the Verification and Validation of Safety-Related Software. *Computers & Standards, 4,* 101 – 112.

Faller, R. (1988). Sicherheitsnachweis für rechnergestützte Steuerungen. *Automatisierungstechnische Praxis atp, 30,* 508 – 516.

Geisselhardt, W. (1978). *Fehlerdiagnose in Geräten der Digitaltechnik.* Carl Hanser Verlag, München Wien.

Görke, W. (1973). *Fehlerdiagnose digitaler Schaltungen.* B. G. Teubner, Stuttgart.

Grimm, K. (1985). Klassifizierung und Bewertung von Software-Verifikationsverfahren. In *Technische Zuverlässigkeit — Generalthema: Softwarequalität und Systemzuverlässigkeit.* VDE-Verlag, Berlin Offenbach. pp. 79 – 90.

Hölscher, H., and J. Rader (1984). *Mikrocomputer in der Sicherheitstechnik.* Verlag TÜV Rheinland, Köln.

IEC (1986). Standard 880 *Software for computers in the safety systems of nuclear power stations.*

Krebs, H. (1984). Zum Problem des Entwurfs und der Prüfung sicherheitsrelevanter Software. *Regelungstechnische Praxis rtp, 26,* 28 – 33.

Krebs, H., and U. Haspel (1984). Ein Verfahren zur Software-Verifikation. *Regelungstechnische Praxis rtp, 26,* 73 – 78.

O'Neill, I. M., D. L. Clutterbuck, P. F. Farrow, P. G. Summers, and W. C. Dolman (1988). The Formal Verification of Safety-Critical Assembly Code. In W. D. Ehrenberger (Ed.). *Safety of Computer Control Systems 1988.* IFAC Proceedings Series, 1988, No. 16. Pergamon Press, Oxford. pp. 115 – 120.

Redmill, F. J. (Ed.) (1988). *Dependability of Critical Computer Systems – 1.* Guidelines produced by The European Workshop on Industrial Computer Systems Technical Committee 7. Elsevier, Amsterdam New York.

Redmill, F. J. (Ed.) (1989). *Dependability of Critical Computer Systems – 2.* Guidelines produced by The European Workshop on Industrial Computer Systems Technical Committee 7. Elsevier, Amsterdam New York.

Reinert, D. (1979). *Prüftheorie diskreter Systeme.* VEB Technik, Berlin.

Schmidt, K. P. (1988). *Rahmenprüfplan für Software.* Formblätter und Anleitung für Prüfungen von Software nach den Güte- und Prüfbestimmungen Software RAL-GZ 901 und der Vornorm DIN 66285 "Anwendungssoftware, Prüfgrundsätze". Arbeitspapiere der GMD 312. Gesellschaft für Mathematik und Datenverarbeitung GmbH, Sankt Augustin.

Xilinx, Inc. (1989). *The Programmable Gate Array Data Book.* San Jose, CA.

SOME ANSWERS TO THE PROBLEM OF SAFETY IN MICROPROCESSOR-BASED DEVICES

J. Ph. Gérardin, C. Vigneron and Ph. Charpentier

National Research and Safety Institute (INRS), Vandoeuvre, France

Abstract. What confidence can we have in a microprocessor based control or monitoring device when the safety of people or the reliability of an industrial process is a stake ?

In order to provide some objective elements in answer to this delicate question, INRS carried out a series of tests on a dozen electronic microprocessor based devices which are widely used in industry.

The tests were carried out using the DEF.Injector device which allows the operating safety of the equipment to be measured.

The results show that certain devices designed for specific applications have a high level of safety while others including the programmable controllers do not guarantee a sufficient safety level in applications where risks are involved.

Keywords. Programmable controllers ; Microprocessors ; Safety ; Microprocessor control ; Perturbation techniques ; Redundancy ; Fault injection ; Validation

"Programmable controllers have never been so safe".

"As far as safety is concerned, it is dangerous to say the least to rely on the programmable controller alone".

Faced with such contradictory assertions the industrial manager hesistates : "Can I safely make the decision to automate my process using programmable systems without creating risks for the people who have to use it ?".

The difficulty arises from the fact that the arguments put forward by the protagonists of these opposing positions are essentially qualitative, even subjective, and therefore are susceptible to contradiction. Statistics to support either position are lacking, or if they do exist are either confidential or too imprecise to be used. The question remains : is it possible to rely on programmable controllers, based on microprocessors, to operate machines, processes, or to protect goods, equipment, or even human life ?

INRS therefore considered it worthwhile to carry out a study to obtain numerical results on the reliability and safety of microprocessor based systems.

This article presents the results of part of the study.

PURPOSE

Our purpose was to evaluate the level of reliability and safety of several microprocessor based electronic devices used in industry. More specifically, we intended to analyse the behaviour of these systems in the event of failures, resulting from internal faults (component failure, design fault) or faults caused by external factors (electromagnetic interference, inductive interference, abnormally high temperature). From the behaviour observed, it is possible to deduce the consequences for the machine or process controlled by the microprocessor based system, as well as the eventual risks run by the operators.

METHODOLOGY

We quickly realised that it was almost impossible to obtain statistical information from most of the designers or users of such equipment. We decided to get this information from laboratory tests.

A series of tests on the electromagnetic susceptibility of programmable controllers had already been carried out at INRS five years ago. This enabled us to determine the level of interference they could tolerate. The tests did not provide information on their level of safety as such. In effect, when the immunity level of the controller was exceeded, it was only possible to specify that an incorrect operation had occurred, but it was not possible to identify what failures had been induced within the equipment, nor to analyse with precision the behaviour of the controller in the face of these failures. A test for electromagnetic susceptibility is not a measure of the level of reliability or safety.

One way of finding out precisely what these two levels are is to measure two parameters called "Coverage Rate" (CR) and "Malfunction Rate" (MR) defined as follows :

The "coverage rate" is the ratio expressed (as a percentage) between the number of faults detected (Ndd) by the system being analysed and the number of faults in total (Ndt) which can arise.

$$CR \% = (Ndd/Ndt) \times 100$$

The "malfunction rate" is the ratio (as a percentage) between the number of serious or dangerous faults (Ndg) and the total number of faults (Ndt) which can arise

$$MR \% = (Ndg/Ndt) \times 100$$

By fault detected we mean an internal fault, or interference which either causes the device to go into a safe state (holding the outputs in a pre-determined state, for example) or the operation of an alarm signal which can be used to prevent an incident arising (watchdog signal for example).

A serious or dangerous fault is one which is not detected and could lead to a hazard to people or the machine. This could involve the issuing of aberrant controls to the process or the suspension of a safety function, such as the emergency stop.

Even so, the total number of faults is difficult to measure. It is impossible to go through exhaustively all the faults which can arise in a complex device, of whatever type, an even more difficult with a microprocessor based system.

It is not possible to calculate the exact values of CR and MR, but it is possible to get approximate values, limited to faults which can be simulated or injected directly into the equipment. Theses faults constitute what are known as fault models.

Our purpose is not to argue the theoretical approach to measuring operational safety but we suggest that the reader consults specialist articles for these aspects (Arlat, 1989).

TEST INSTRUMENT

The objective of our tests was to determine the two parameters CR and MR of several industrial and experimental devices. We used test instrument specially adapted for this purpose : the DEF-Injector also called DEF.I (Gerardin, 1989).

Developed by INRS, this instrument, now produced by SCHAFFNER provides designers with a way of measuring the safety performances of their products.

With DEF-Injector faults can be simulated within the equipment under test (target equipment) and its behaviour in the presence of these faults analysed in real time.

The instrument offers a choice of fault model : memory faults, microprocessor faults, faults in the bus. It also allows a choice of the nature of simulated fault (permanent or transient fault) as well as the time of their appearance in the target device (at start up, or during operation). A permanent fault represents a complete failure of a component, whereas a transient fault reflects the consequence of interference or a variation in operation.

For each simulated fault, DEF.I analyses the reaction of the target.

DEF.I allows several dozens or hundreds of thousands of faults to be simulated, without risk of damage.

DEF.I identifies faults not detected by the target equipment, and also serious faults, and calculates the coverage rate and malfunction rate.

FIELD OF INVESTIGATION

Tests were carried out on 11 systems :

Two dedicated devices designed specially for the control and monitoring of a specific process. One was a control and monitoring device for a nuclear power station and one a burner control and safety device.

Two programmable controllers from the middle of the range of two different manufacturers ;

Three remote transmission devices - two industrial remote controllers and one alarm device for the protection of isolated workers ;

Two safety devices - a light curtain and a controller for several photoelectric cells providing zone protection ;

Two experimental devices designed to test the methods of ensuring the safety integrity of microprocessor based systems. The first is a device whose safety is based on signature analysis of the instructions carried out by the microprocessor (Schweitzer, 1988). The second uses redundancy - two identical central processing units execute the same program at a slight difference in time. The two "Data Bus" are compared to ensure they carry out identical functions (Charpentier, 1988).

TESTS

The 11 devices were subjected to a series of tests with DEF.I, varying the test parameters : fault model (memory faults, microprocessor faults) ; nature of faults (permanent or transient) ; moment when the faults arises (start up, during operation).

Determining the coverage rate

The information indicating the fault had been detected by the target device, analysed by DEF.I in order to calculate the coverage rate, appeared under different forms according to the device :

Control and monitoring device of a nuclear power station : safety output to a majority voter monitoring three elements identical to the one tested ;

Burner control and safety device: interlocking of a safety relay preventing the activation of various output relays controlling the main valve and ignition ;

Programmable controllers : state of alarm relay used to cut power to the outputs in case of failure of the controller ;

Industrial remote controllers : state of relay controlling the cutting off of a group of output relays governing emission of orders ;

Alarm system : state of a fault indicator ;

Safety devices : state of output relay intended to stop the process in the event of access into the protected zone ;

Experimental devices : output alarm signifying a signature not in conformity with that expected or output from the comparison system indicating discordance between the two CPUs.

Determination of the malfunction rate

It is not possible to determine this parameter unless the behaviour of the device which could lead to a hazard arising from its application is clearly identified. The dangerous behaviour which we have identified for the equipment analysed are the following :

Burner control and safety device : failure of safety function in the event of flame failure ;

Industrial remote controllers : operation of an output relay not corresponding to a command ;

Alarm system : failure to register an alarm ;

Safety devices : failure to open the output relay which stops the process in the event of access into the protected zone ;

Programmable controllers : the measure of the safety level of a controller as such does not make much sense. It has no significance except for a controller programmed for a particular application whose hazards can be identified. We have therefore measured the malfunction rate of one of these controllers, having programmed it to control and monitor a gas burner in the same sequence as the specialized device we have also analysed. The hazardous situation is the same as for the above device, failure of the safety function following loss of flame ;

Experimental devices : not having been designed for a particular application, we have not tried to measure their malfunction rate ;

Control and monitoring device for a nuclear power station : this device being a prototype and not integrated in the complete system, this test was not attempted.

RESULTS

Because of the automation of the test sequences by the DEF.I, it was possible to introduce tens of millions of faults into the devices being tested. Numerous interesting observations have been made, notably concerning the performance of the various error detection measures in the devices and their sensitivity to the different types of fault simulated.

Complementary studies are being pursued to refine these initial results which will interest primarily the designers of microprocessor based electronic equipment as well as those who evaluate their performance experimentally, or for certification purposes.

Our objective was to isolate from these tests objective and quantifiable results capable of informing the users of such equipment about the confidence they can have in them, and hence we present here only the results concerning this aspect. These are given in Tables 1 and 2. These give the values of the two parameters measured for a given fault model in the various devices analysed. The results given here are those corresponding to a model in which each fault simulated manifests itself by a unique error in the programm executed by the microprocessor in the target device ; this error can be permanent or transient. The results have been obtained from a number of faults simulated in the target, varying from tens of thousands to several hundred thousands, according to the program size in the device.

We have separated industrial devices (Table 1) from experimental devices and the prototype device for a nuclear power station (Table 2).

In these tables the devices are ranked according to decreasing value of coverage rate (for permanent faults).

Table 1

Device tested	Coverage rate		Malfunction rate	
	permanent fault	transient fault	permanent fault	transient fault
Burner control and safety device	68 %	45 %	0,002 %	0 %
Central controller for photoelectric	68 %	48 %	0,06 %	0,001 %
Light curtain	64 %	49 %	3 %	0 %
Remote controller x	60 %	10 %	1,5 %	0,05 %
Remote controller y	60 %	17 %	1,8 %	0,02 %
Alarm system for isolated workers	30 %	17 %	20 %	2 %
Programmable controller x	27 %	20 %	-	-
Programmable controller y	10 %	5 %	37 %	7 %

Table 2

Device tested	Coverage rate	
	Permanent fault	Transient fault
Control and monitoring unit for nuclear power stations	99,95 %	-
PES with double redundancy of hardware and software operating asynchronously	99,7 %	99,6 %
Single channel system with signature analysis	95,8 %	95,4 %

SIGNIFICANCE OF RESULTS

To analyse the informations which can be obtained from these results, let us examine first of all the parameter coverage rate, and then the parameter malfunction rate.

Coverage Rate

Coverage rate considered as a feature of quality : The coverage rate can be considered as a measure of the quality of a device. This parameter quantifies the effectiveness of the mechanisms for detecting abnormal operation which are (eventually !) integrated into the system. It permits an assessment both of the importance given by the designer to the criterion of reliability and of his ability to realise it. It therefore constitutes an important criterion in the choice of equipment in circumstances when dependability is particularly important and it must be said that there are few areas where dependability is a secondary matter. In effect, a definition of dependability is the quality of service delivered by the system, quality in which the user can be justly confident. Seen from this angle, we think that few manufacturer would be prepared to admit that they don't think it is important.

This is why, between two devices capable of fulfilling a range of duties (capacity, reliability, cost, etc) a manufacturer will always benefit from choosing the one which offers the best coverage rate. The higher the coverage rate the more likely it is that an equipment failure which could affect the application (tool damage, loss of production) is limited. In this sense, controller x is more to be recommended than controller y.

The results of these tests sometimes surprised us in that they contradicted the first impression we had of the dependability of a device following a quick examination of its hardware and software. We initially thought that remote controller y would be superior to remote controller x in respect of operating safety. The design and construction of the first seem to be more "serious" than the second. The result of these tests showed this impression to be wrong because the two devices have very similar coverage rate values.

So, it must be said that only a true validation test like the one we have carried out will be capable of evaluating the dependability performance of a system with any precision. In this field beware of judging by "outward appearances".

A further example to confirm this concerns the two programmable controllers. Their error detection system was based on the use of a similar watchdog. The tests nevertheless showed that controller x is nearly three times more effective at error detection than controller y (coverage rate respectively 27 % and 10 %). A detailed analysis has shown that this is due to a design fault in the watchdog of controller y.

Reading the tables, it can be seen that the coverage rate for the detection of transient faults is always less than that for permanent faults. This could be because the detection mechanisms are less able to identify errors of this type, or it could be that the designer ignores transitory errors so as not to limit the availability of his equipment (ie if it shuts down too often in response to transitory errors it will be less available for production control and monitoring). It is not possible to draw immediate conclusions concerning the performance of the device analysed. However, since transient errors are most likely to occur during use, it is necessary to study their effect on safety carefully, analysing the corresponding malfunction rate, as we do later on.

The coverage rate shown in Table 2 are all greater than 95 %. This shows that it is possible to achieve a very high level of reliability with microprocessor based systems when this critirion has been the main objective of the designer.

Coverage rate and complexity of the system:
Table 1 shows that the coverage rate values range from 10% to 68%.

It would be incorrect to deduce, for example, that the ability of the manufacturers of the remote controllers we tested is six times greater than that of the manufacturer of the programmable controller y.

The complexity of the system must be taken into account.
The more complex the system, the more difficult it is for its designer to take account of all the variables which could affect its operation.

For the equipment we have analysed, the complexity can be evaluated as a function of the quantity and the nature of inputs the system receives from the exterior. Three categories of equipment, of increasing complexity, can be distinguished :

-   systems which operate with little information from outside : this is the case with the two safety devices and the burner control and safety device where the information is less than six binary inputs ;

-   systems receiving more information than this, but where this information is not strictly independent. This is the case of the remote controllers and the alarm system,

all of which use strictly coded messages incorporating redundancy of information ;

-   systems which handle a large quantity of information of diverse types. In this category are the programmable controllers, the control and monitoring device for the nuclear power station and the two experimental devices.

The results of Table 1 reflect this classification. It is only within these three categories that one can draw conclusions on the ability of the manufacturer. So, it is possible to say that the three devices in the first category come from manufacturers who have taken into account the criterion of credibility. By contrast, in the two other categories there are clear differences in performance. The manufacturer of the alarm system has clearly not taken account of the consequence of an internal failure on the operation of his system (CR of 30% compared with 60% for the remote controllers). As for the manufacturer of controller y, we have already seen that the poor performance of this product derives from a design fault.

The results in Table 2 show that complexity in the sense in which we have defined it, and reliability are not opposites : it is possible to get a high coverage rate value for systems which are relatively complex.
INRS's researchs are particularly concerned with obtaining a good correspondence between performance and cost.

Malfunction Rate

If safety is an important factor in a given application, determination of the coverage rate will not be sufficient. It will be necessary to know the effect on the operation of the system of errors which escape detection. This is provided by the measure of the malfunction rate.

Table 1 shows that two systems which have an equivalent level of detection of errors can differ as far as the safety criterion is concerned. The two safety devices we tested have levels of safety which differ greatly despite the similar coverage rate values (64% and 68%) : for the light curtain, 3% of the permanent faults simulated made it blind (interrupting the infra red beam did not cause the output relay to drop out) whereas the rate was only 6 in 10,000 for the central controller for the photoelectric cells.

This shows that it is essential to validate by fault simulation to make correct assessment of the level of safety of a device.
A second point is the difference in behaviour of the systems analysed according to the nature of faults simulated.

The equipment was always affected in a less hazardous way by transient faults than by permanent faults.

This point is the more significant since the sensitivity of microprocessor based systems to transient faults (surges, interference) is something generally seen as being incompatible with an acceptable level of safety integrity. We are wary of drawing hasty conclusions, our sample of equipment tested being too limited to put this forward as a general rule. Nevertheless, we will find out if the phenomenon is confirmed or invalidated in our subsequent studies.

The most interesting fact arising from the analysis of the malfunction rate is the enormous difference in performance between the burner controller and safety device and the performance of the same function by controller y.

In the first case, only one fault in 56 000 could, in certain test conditions, inhibit the safety shutdown of the microprocessor based control following loss of flame, whereas in the second case more than one fault in three could lead to this dangerous situation !

From this we conclude :

Firstly, that it is possible to find microprocessor based devices with a high level of safety integrity and costing in the same region as similar devices which do not take into account the safety criterion.

Secondly, that it appears dangerous to use general purpose devices (controllers, calculators...) for which the safety criterion was not the prime objective of the designer in applications where this parameter is important.

CONCLUSIONS

We believe that the results obtained during this series of tests are a basis for responding in an objective way to the question "What level of confidence can be given to microprocessor based systems used in industry ?".
The tests have proved that it would be ill considered, to say the least, to afford the same level of confidence to all systems. At the present state of the art, only devices specifically designed for a specific operation can achieve an acceptable level of safety integrity. This level cannot truly be determined except as a result of validation test using fault injection or simulation. Whenever safety is an important component of an application, this type of validation is indispensable.

It is therefore up to the user to demand of the suppliers or from his research department the results of such a procedure before allowing a device to be incorporated in an application where the risks from the process or for the operators could be serious.
It is particularly important not to use general purpose devices (programmable controllers, micro computers, standard circuit boards) for this type of application.

When the use of these systems cannot be avoided completely, the safety function should be performed by specifically designed and validated complementary devices .

REFERENCES

Arlat, J., Crouzet,Y. and Laprie, J.C. (1989).
Fault-injection for dependability validation of fault - tolerant computing systems
19th International Symposium on Fault-Tolerant Computing (FT CS 19) Chicago (USA) 21-23 juin 1989, ed. IEEE, 348-355.

Charpentier, P., Kopka, B. and Gérardin, J.P. (1988).
Etude et validation d'un système à microprocesseur à haut niveau de sécurité basé sur une redondance à hétérogénéité temporelle. 6ème colloque international de fiabilité et de maintenabilité. Strasbourg, 3-7 Octobre 1988.

Gerardin, J.P. (1989).
The "DEF. Injector" Test Instrument, Assistance in the Design of Reliable and Safe Systems
North-Holland Computers in Industry 11, 311-319.

Laprie, J.C. (1989).
Dependability of resilient computing systems
T. Anderson editor, Blackwell Scientific Publications, 1-28.

Schweitzer, A. (1988).
Improving the safety level of programmable electronic systems by applying the concept of signature analyses. Safety and Reliability of Programmable Electronic Systems, Edited by B. K DANIELS. Elsever Applied Sciences Publishers, 199-209.

KNOWLEDGE MODELLING AND RELIABILITY PROCESSING: PRESENTATION OF THE FIGARO LANGUAGE AND ASSOCIATED TOOLS

M. Bouissou, H. Bouhadana, M. Bannelier and N. Villatte

Electricité de France, DER/ESF Section,
1 av. de Général du Gaulle, 92141 Clamart cedex, France

Abstract. EDF has been developing for several years an integrated set of knowledge-based and algorithmic tools for automation of reliability assessment of complex (especially sequential) systems.

In this environment, the reliability expert has at his disposal all the powerful classic tools for qualitative and quantitative processing and besides he gets various means to generate automatically the entries for these tools, through the acquisition of graphical data.

The development of these tools has been based on FIGARO, a language for system modelling, which plays an important unifying role.

A variety of compilers and translators transform a FIGARO model into conventional models, such as fault-trees, Markov chains, Petri Nets...

In this paper, we present the main ideas which determined the FIGARO language, and we illustrate these general ideas by examples.

Keywords. Knowledge representation, Modeling, Simulation, Stochastic systems, Reliability, Performance, Monte Carlo methods, Markov chain.

I. INTRODUCTION

In the framework of the probabilistic safety analysis of the Paluel nuclear power plant (EPS 1300), EDF has developed software packages allowing the automation of reliability models' construction and assessment.

These tools were used to develop new concepts, original and highly performing algorithms /1/, but they lacked generality and user friendliness. The main problem lay in the fact that the expert systems being applied for generation of reliability models were too specific of the fields being dealt with and difficult to maintain.

EDF has therefore developed a second generation of these software packages. This version, which is available on a workstation (under UNIX/Xwindow) with user friendly, graphical interfaces, is no longer dedicated to nuclear applications.

Our concern for unification of the software packages, explanation of the reliability expert's modelling choices, and generality has led us to design a unique system modelling language (the FIGARO language) which is independent from the processing method used afterwards. This language has been worked out in order /4/:

- *to give a suitable formalism for setting up knowledge bases* (with generic component descriptions),
- *to be more general* than all conventional reliability models,

- to make the best possible *compromise* between *modelling power* (or generality) and processing *tractability*,
- to be as *readable* as possible,
- to be easily associated with graphic representations.

In fact the setting up of knowledge-based systems is the only way to reduce significantly the necessary outlay for the reliability studies.

On the basis of a FIGARO language modelling, different compilers and translators allow to deduce automatically the data which are necessary for the classical reliability model processing codes: fault trees, Markov chains, Petri nets, etc.

II. THE FIGARO LANGUAGE MAIN OBJECTIVE: TO MODEL DISCRETE STATE SPACE SYSTEMS

Let's take a physical system. We can define three probability model categories. Starting from the most detailed (and most complex) models to the least detailed ones, the specified categories are as follows:

A. Continuous state space dynamic simulation models,
B. Discrete state space dynamic simulation models,
C. Abstract models.

A model of type A is made up of:

- The deterministic differential equations which rule the system physical quantities: temperature, pressure, mass, etc...
- The discontinuities due to sudden component state changes induced by random phenomena (faults, repairs...) or deterministic (timer triggered action...).

A model of type B does not imply differential equations: the system can only have a finite or countable number of states and runs over from a state to another following a random or deterministic phenomenon.

The evolution of the system is therefore a *continuous time stochastic process* which can be represented schematically as follows:

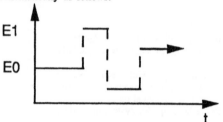

The models of type C (for example : fault trees) are thoroughly different from the two preceding ones as they have only a remote relation with the physical phenomena which rule the system life. The time is not introduced explicitly. That's why we have called them "abstract" as opposed to the simulation models.

We wished to find, in the definition of a language which describes probabilistic problems, a fair equilibrium between "full" simulation, which is quite detailed, and a too abstract model.

It is obvious that the first type of model, which is ideal in the absolute, is in practice too rich to be tractable (in most cases).

On the contrary, the choice of the C-type model (for instance a fault tree), or even of a too restrictive B-type model (for instance a Markov chain) may lead to unacceptable simplifications.

Therefore we have tried to work out a language which could describe unambiguously a B-type model, keeping in mind the objectives given in part I.

An analysis of the existing modelling and computer languages has shown that none of them provided the required set of characteristics; therefore we have created the FIGARO language with specific object-oriented type syntax and semantics. FIGARO is part of the so-called "hybrid" languages, that is to say it takes some of its features from the object-oriented languages and models the behavior of an object through order 1 production rules.

The use of rules offers two advantages:

- The rules are close to the natural language if their syntax is selected appropriately: their use will improve the model readability,

- EDF has got the mastery of different tools in this field and, in particular, worked on the validation of 0 order production rule bases.

The characteristics which make of FIGARO an object-oriented language bring some decisive advantages in *knowledge storage* and among them:

- Easy knowledge classification,

- No repetition (due to the "factorization" allowed by the heritage mechanism) which is a source of maintenance errors.

A FIGARO description is made of two kinds of rules:

- interaction rules: they model the propagation of instantaneous effects.
  Ex : IF (flow(pump1) + flow(pump3)) < threshold2
      THEN state(alarm) <- 'on';

- occurrence rules: they yield the list of events that may happen in a state of the system. These rules have a particular semantics, related to time; this is why they include a distribution law of the time after which the event will happen:
  Ex1 : IF state(timer) = 'on'
      THEN-MAY-HAPPEN
      EVENT down
      EFFECT state(timer) <-- 'off'
      LAW FIXED_TIME(delay) ;
      (* deterministic law *)

  Ex2 : IF state(engine) = 'working'
      THEN-MAY-HAPPEN
      FAILURE breakdown
      EFFECT state(engine) <-- 'breakdown'
      LAW EXPONENTIAL(lambda) ;
      (* constant failure rate *)

Important note: it can be easily proved /4/ (and that will appear in the below-mentioned examples) that FIGARO has the very important property to be a generalization of most of the modelling types used by the reliability experts; it generalizes in particular:

- the fault trees,
- the state graphs (including Markov chains),
- the stochastic Petri nets,
- the queuing models.

This property of FIGARO ensures that any conventional model has its FIGARO equivalent and therefore the exclusive use of FIGARO as representation formalism is not restrictive whatsoever.

III. FIGARO MODELS PROCESSING METHODS

The processing operations take place in two stages in order to master the combinatorial explosion problems at best:

*The first is only intended to pass over from the FIGARO representation of order 1, made up of the knowledge base types and objects describing a particular system, to an exactly equivalent FIGARO representation of order 0 which is obtained through application of the heritage mechanism to the objects and by instanciation of the first order (generic) rules in the form of zero order (specific) rules.

For example, the following first order rule (in the type "circuit-breaker"):

IF position = 'open' OR (FOR_ANY x AN upstream_component, energised(x) = FALSE) THEN FOR_ALL y A downstream_component DO energised(y) <-- FALSE ;

will produce this much simpler zero order rule, for a circuit breaker cb1, having the two u1 and u2 upstream components, and the d1 downstream component:

IF position(cb1) = 'open' OR (energised(u1) = FALSE AND energised(u2) = FALSE) THEN energised(d1) <-- FALSE ;

The transformation of first order rules into zero order rules is interesting for two reasons:

- direct processing from the first order would be too difficult as it would require numerous evaluations of first order rules which are time consuming,
- the rule base coherence checking tools exist only for zero order rules.

*The second is the choice (more or less automatized) of the processing method and the application of this method.

In order to choose the method, one has to determine the more or less static character of the system, in other words a more or less great independence between different parts of the model.

When it is possible to identify independent parts, one should take advantage of this feature to study these parts separately.

The breakdown of a big model into sub-models is a fundamental heuristic which offers decisive advantages:

- the sub-models are simpler to understand, to validate,

- they imply easier processing (the memory and the CPU time required for a study are exponential functions of the model size).

- due to their characteristics, the sub-models allow processing operations which are impossible for global models (example: a sub-model can be Markovian but that is not the case of the global model).

In order to achieve that, we have developed a program which elaborates the influence graph between the state variables of the FIGARO 0 model and allows different processing operations on the basis of this graph.

With this software, the user works in an interactive way: he can display, on request, different data on the dependence graph and order the output of sub-models (extracts from the global model in FIGARO 0) on files which will be reinserted at the start of the processing chain in order to be translated into quantifiable models.

When the user has identified the FIGARO O pertinent sub-models, he can choose the optimum method for each of them and generate the input data of one of the available evaluation tools by the adequate translator.

The evaluation tools now used at ESF and fully integrated in the FIGARO environment are as follows:

- for any FIGARO model: GSI, a software which has been developed since 1985 at ESF. GSI offers a wide variety of processings, all available from a single description (in zero order production rules). These processings are listed in part IV.

- for fault trees: PHAMISS /6/, a Dutch software developed by ECN. This software is remarkable through the variety of the quantifications which it can carry out: calculation of availability, reliability, importance of components and minimal cut sets, uncertainty and this is for components being repairable or not, periodically checked or not. The automatically generated trees have characteristics (such as many repeated leaves) which cause PHAMISS to lose much time: that's why we have developed a tree optimizer which carries out a pre-processing /2/.

- for stochastic Petri nets: either GSI, or MOCA-RP /5/, a code developed by ELF-Aquitaine which allows to carry out a Monte Carlo simulation.

IV THE GSI EVALUATION TOOL

The input of GSI is a rule model, including (like FIGARO) occurrence and interaction rules. All the rules are specific (like in FIGARO 0).

GSI allows three main types of treatments, corresponding to different methods:

- **Monte Carlo simulation**: this method is the most general, and can be used even for non Markovian models, including various lifetime distributions for the components (lognormal, Weibull...), and deterministic phenomena (fixed time type laws). The simulation can yield virtually any kind of result about the behavior of the system: reliability and availability, of course, but also average performance, number of events, sojourn times... The only limit of this method is well known : it may be very time consuming.

- **Markov chain generation and quantification**: this method is applicable when the model contains only exponential and instantaneous probability laws, in other words, when the model describes a Markov stochastic process, with a finite number of states. The (very severe) limit of this method is the exponential growth of the number of states when the size of the system increases.

- **Sequence generation and quantification**: this last method is the most original, and has given its name to GSI: "Generation of Sequences by Inferences" /3/, /7/.

This approach does not require graph production and allows to deal with models which are equivalent to huge, even **infinite**, Markov chains.

As a matter of fact, GSI, thanks to different heuristics, allows to limit the number of sequences to explore: to each type of exploration corresponds an approximation level.

In 89-90, theoretical works run in cooperation with the ORSAY university have permitted, besides showing that the GSI quantification techniques remain valid on infinite graphs, to get better estimations of the errors due to the approximations.

Another advantage of the sequence generation method is that it gives the most probable sequences, i.e., the weak points of the system.

Since all the treatments of GSI may be very time consuming for big models, this software works in two steps: the input model is first translated into PASCAL procedures, which are compiled and linked with the fixed procedures containing the algorithms; then the execution itself takes place.
This technique is necessary to be able to run models containing thousands of rules in reasonable times.

V. APPLICATION FOR A USER FRIENDLY WORK ON CONVENTIONAL RELIABILITY MODELS

To carry out quick and simple studies or such studies which do not justify the development of reusable FIGARO component descriptions, the reliability expert may want to use a conventional model, such as a fault tree, a reliability diagram, a Petri net...

Such a model will be built more rapidly than a FIGARO based model, which obliges to structure and formalize the concepts being handled more or less consciously in the production of the specific model. In return, it won't be at all reusable for carrying out a second study of the same type.

FIGARO gives a very satisfying answer to this request through the possibility it offers to create "knowledge micro-bases" corresponding to the classical models. These bases allow *graphic model acquisition.*

Their development is quite fast (in general one day).

Therefore this allowed us to create easily a coherent set of graphic interfaces (as it rests on a single tool) for all the conventional models.

More generally, it is important to notice that any simple graphic language can be supported by means of a small, quickly written FIGARO knowledge base.
Besides, the FIGARO based modelling allows to access the full available processing set: for example it is possible to assess the *reliability* of a system represented through a fault tree by a Monte Carlo simulation, which is feasible whatever the FIGARO model, or by the analytical calculations of GSI whereas most of the fault tree codes do not permit such a calculation (for a repairable system).

Fig. 1 and Fig. 2 at the end of this paper show the aspect of graphic interfaces, which are set up through FIGARO for an example of reliability diagram, and of Petri net.

The reliability diagram can be calculated on request by PHAMISS after transformation into a fault tree, or directly by GSI.

The Petri net, as far as it is concerned, can be assessed on request by GSI, or by MOCA-RP.

VI. APPLICATION FOR STUDY OF SEQUENTIAL ELECTRICAL SYSTEMS

A knowledge base has been developed in order to model and quickly evaluate the reliability of nuclear power plant electrical distribution systems.
This knowledge base includes components such as: diesel-generators, busbars, circuit-breakers, transformers...
All the sequential aspects of this kind of systems (automatic reconfigurations after failures and repairs) are explicitly modelled in this knowledge base.
Fig. 4 gives examples of sequences which lead to the failure of the system of Fig. 3. GSI automatically found these sequences, *without building the underlying Markov chain.*

VII. APPLICATION FOR COMMUNICATION NETWORKS

A knowledge base was written to allow the quick comparison of different topologies of a communication network. The network is supposed to be made of nodes and links, which both may have failures; all the components are independent.
After the acquisition of the topology of a network, it is possible to study it either by generating a fault-tree, or by running GSI.

In a second knowledge base, a refinement has been introduced in the modelling: it is possible to declare that several components share the same repairmen. This introduces dependencies into the system, and the fault tree study remains no longer valid, whereas GSI can still be used.

VIII. DEVELOPMENT TOOLS, TARGET MACHINES

The Table 1 sums up, on a practical basis, the main features of the tools which have been quoted in this paper.

IX. CONCLUSION

We have provided the main characteristics of the FIGARO application prototype by illustrating through examples its high generality degree and user friendliness.

This application makes up a complete environment in which:

- a user of knowledge bases can carry out "fully mouse-controlled" system studies rapidly enough (10 to 20 times faster than without knowledge base) to compare different system design solutions,
- the developer of knowledge bases has at his disposal a high level formalism (the FIGARO language) to

express the (functional or material) knowledge he wants to formalize.

For the time being, the user can still take many initiatives in processing selection from a FIGARO model and the developer is totally free in writing a knowledge base. In particular he can choose the modelling fineness: FIGARO allows reliability processing from a very simplified component modelling. In this way, the inexperienced user can gradually pass to the knowledge base developer's stage through gradual improvement of his models.

The following stage consists in obtaining, through an intensive use of these tools in fields as different as possible, more directing guides in order to help:

- the developer to structure and formalize his knowledge when he builds a knowledge base,

- the user to choose the most pertinent processing, in particular by controlling the validity of the choices he makes.

REFERENCES

/1/ Ancelin, C., Bannelier, M., Bouhadana, H., Bouissou, M., Lucas, J.Y., Magne, L., Villatte, N. (1990). Poste de travail basé sur l'intelligence artificielle pour les études de fiabilité.
Revue Française de Mécanique, Numéro spécial : les systèmes experts et la mécanique N° 1990-2 . ISSN 0373-6601.

/2/ Bouhadana, H. (1989). Méthodes d'amélioration qualitative d'un modèle de fiabilité.
Mémoire de DEA. EDF/Paris XIII. Sept. 1989.

/3/ Bouissou, M. (1986). Recherche et quantification automatiques de séquences accidentelles pour un système réparable.
5ème Congrès de Fiabilité et Maintenabilité de Biarritz (France). Oct. 1986.

/4/ Bouissou, M., Bouhadana, H.,Bannelier, M. (1990). Un moyen d'unifier diverses modélisation pour les études probabilistes :
le langage FIGARO.
HT-53/90-42A. EDF internal report.

/5/ Signoret, J.P. (1989). MOCA-RP batch, utilisation du logiciel, révision 0.
Internal report SNEA(P) DEA-SES-ARF, JPS/cl/n°89-71.

/6/ Terpstra, K., Dekker, N.H., Van Driel,G. (1986). PHAMISS, A Reliability Computer Program for phased Mission Analysis and Risk Analysis. User's Manual.
ECN-83. Netherlands Energy Research Foundation.

/7/ Villemeur, A., Bouissou, M., Dubreuil-Chambardel, A. (1987). Accident sequences: methods to compute probabilities.
International topical conference on PSA and Risk Management. (Zurich, Switzerland).

TABLE 1 : development tools and target machines

	Programming language	Machines and systems	Developer	AI techniques.
FIGARO language definition	------------	------------	EDF	Knowledge representation. Object language. Order 1 production rules.
Graphical Interface	LELISP AIDA	SUN 3 and 4 (UNIX/X11)	EDF	Object language.
Compilers (FIG0, GSI, fault tree)	LEX/YACC C LELISP	SUN 3 and 4 (UNIX)	EDF	Backward chaining (fault trees). Order 1 to order 0 rule transformation.
GSI V5.3	PASCAL VS PASCAL ISO	IBM 3090 (MVS), SUN 4, HP 9000 (UNIX), IBM PS (AIX)	EDF	Inference engine in forward chaining. Heuristic rules. Compilation of 0 order rules.
Dependence analysis	LEX/YACC C	SUN 4	EDF	-------------------
PHAMISS	FORTRAN 77	IBM 3090 (MVS), SUN 4 (UNIX), IBM PS (AIX)	ECN	-------------------
MOCA-RP	FORTRAN 77	IBM 3090 (MVS), SUN 4 (UNIX), IBM PC (MS.DOS)	ELF Aquitaine	-------------------

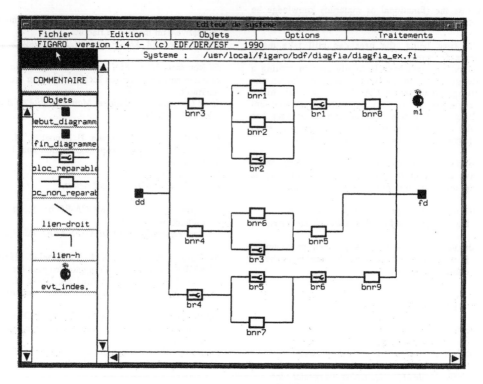

Fig. 1 : Graphic data acquisition for a reliability diagram

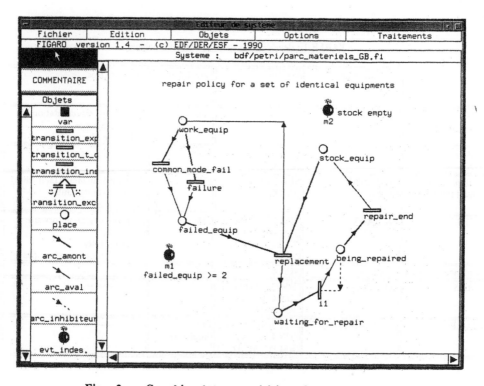

Fig. 2 : Graphic data acquisition for a Petri net

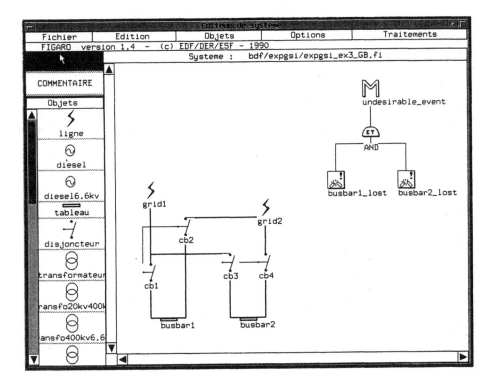

Fig. 3 : Graphic data acquisition for an electrical system

```
****************************************************************
* SEQ *        EVENTS        * REPAIR * DURATION *ASYMP PROB* PROBA/T1 *
* NUM. *                     * TIME   *          *          *          *
****************************************************************
*    4: loss OF grid1        *        *          *          *          *
*       opening OF cb3       *        *          *          *          *
*       opening OF cb1       *        *          *          *          *
*       closing OF cb4       *        *          *          *          *
*;      ref_to_close OF cb2  *        *          *          *          *
*       loss OF grid2        * 9.615e+01* 1.923e+02* 1.704e-08* 4.704e-11*
*    5: loss OF grid1        *        *          *          *          *
*       ref_to_open OF cb3   *        *          *          *          *
*       ref_to_open OF cb1   * 1.503e+00* 0.000e+00* 1.257e-08* 3.468e-11*
*    6: loss OF grid1        *        *          *          *          *
*       opening OF cb3       *        *          *          *          *
*       ref_to_open OF cb1   *        *          *          *          *
*       ref_to_close OF cb4  * 2.983e+00* 0.000e+00* 1.256e-08* 3.467e-11*
*    7: loss OF grid1        *        *          *          *          *
*       ref_to_open OF cb3   *        *          *          *          *
*       opening OF cb1       *        *          *          *          *
*       ref_to_close OF cb2  * 2.983e+00* 0.000e+00* 1.256e-08* 3.467e-11*
*    8: loss OF grid1        *        *          *          *          *
*       opening OF cb3       *        *          *          *          *
*       opening OF cb1       *        *          *          *          *
*       ref_to_close OF cb4  *        *          *          *          *
*       ref_to_close OF cb2  * 1.923e+02* 0.000e+00* 1.256e-08* 3.467e-11*
*    9: loss OF grid2        *        *          *          *          *
*       inadv. open. OF cb1  *        *          *          *          *
*       loss OF grid1        * 9.615e+01* 3.845e+02* 2.507e-09* 6.919e-12*
*   10: loss OF grid2        *        *          *          *          *
*       inadv. open. OF cb3  *        *          *          *          *
*       loss OF grid1        * 9.615e+01* 3.845e+02* 2.507e-09* 6.919e-12*
*   11: loss OF grid1        *        *          *          *          *
*       opening OF cb3       *        *          *          *          *
*       opening OF cb1       *        *          *          *          *
*       ref_to_close OF cb4  *        *          *          *          *
*       closing OF cb2       *        *          *          *          *
*       inadv. open. OF cb2  * 1.923e+02* 1.923e+02* 1.846e-09* 5.096e-12*
*   12: loss OF grid1        *        *          *          *          *
*       opening OF cb3       *        *          *          *          *
*       opening OF cb1       *        *          *          *          *
*       closing OF cb4       *        *          *          *          *
*       ref_to_close OF cb2  *        *          *          *          *
*       inadv. open. OF cb4  * 1.923e+02* 1.923e+02* 1.846e-09* 5.096e-12*
```

Fig. 4 : Sequences found by GSI for the system of Fig. 3.

Fig. 3 : Graphic data acquisition for an electrical system

Fig. 4 : Sequences found by GBI for the system of Fig. 3.

USING FAULT TREE ANALYSIS IN DEVELOPING
RELIABLE SOFTWARE

E. O. Ovstedal

SINTEF DELAB, 7034 Trondheim, Norway

Abstract. The Fault Tree technique can be used to analyse the software developing process.
A Fault Tree is a logic diagram that displays the interrelationships between a potential critical event (top event) in a system and the causes for this event. The goal is to find critical
events concerning fault introduction and to relate those events to the developing process. In
this way we can find appropriate actions that will reduce the probabilities of these events
and the subsequent probability of the top event. An example is used to show an application
of the technique. The analysis of the Fault Tree can be done qualitatively or quantitatively.
The Fault Tree makes it easy to show how the probability of the top event is reduced when
quality assurance activities are introduced into the developing process.
Keywords: software engineering, failure detection, quality assurance, Fault Tree technique

INTRODUCTION

The overall goal for most software developers is to
deliver a software system that satisfies the customers
needs. In reality, most software systems in operation
contain one or more faults or cases of missing functionality.

Missing functionality and faults are introduced at all
stages during the development process. Important
questions are:

- is one part of the development process more
  important concerning reliability than other parts,
  i.e. in which phases are most faults introduced?

- how can we eliminate these faults? In particular:
  how can quality assurance during the development process reduce the probability of
  introducing faults and increase the probability to
  remove the faults?

There are no simple answers to these questions. Each
project is different from other projects. When delivered, the same system will be used differently by
different users. Consequently, we need methods that
can analyse each project and product in order to find
solutions for this particular project or product.

The rest of this paper is organized as follows:

- first: what do we know about fault introduction
  and the probability of their discovery and subsequent removal

- then we will discuss how the Fault Tree technique can be used to identify the important
  phases of the development process concerning
  fault introduction

- at last we will discuss how quality assurance can
  reduce the probability of introducing faults

PROBLEM DESCRIPTION

Definitions and Terminology

We will use the following definitions in accordance
with (IEEE, 1983).

- an error is a human action that insert one or more
  faults in the software

- a fault is a manifestation of an error in the software, i.e. a defect dormant within the system

- a failure is a departure of program operation from
  program requirements. A failure may be produced when a fault is encountered.

Fault avoidance is, according to Laprie (1989),
defined as techniques to prevent, by construction,
fault occurrence and fault introduction.

Fault removal is defined as techniques used to minimise the presence of faults (Laprie, 1989).

Fault detection is defined as techniques used to find
faults in the software.

The Developing Process

There are different developing models for software systems. They all have in common the process model shown in fig.1. The problem to be solved by the software system is often not clearly defined. It is integrated with other problems and with the environments. A cloud is therefore a good symbol for the problem.

The software requirement specification usually reflects the degree of clarity of the problem. It is a mixture of functional requirements, some vague non functional requirements, some political statements and a lot of good hopes and wishes. Missing functionality and inconsistencies are common in the requirement specification. This document is usually the only document given to the system engineers. Based on this he is expected to design, implement and validate the system.

The only model of the system at this stage is the requirement specification. The design can be verified against the requirement specification if it is written in a formal specification language. The implementation on the other hand, can only be validated against the real world problem that the system is supposed to solve.

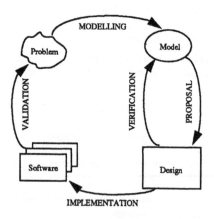

Figure 1

In Which Phases are Faults Introduced

The software developing process is error prone. Faults are introduced in all phases of the developing process. Some experiences on the introduction of faults related to phases are reported in the literature.

Sayet (1990) reports on a project for the developing a safety critical system. He found that of all faults detected 46% were found in the specification phase, 46% were found in the coding phase and 8% were found during the integration test.

Basili (1984) reports similarly results. He found that 48% of the errors were due to incorrect or misinterpreted functional specification or requirements, i.e. errors done in the specification phase of the project.

Our own experience is in accordance with this. Complete and consistent requirement and functional specifications are of fundamental importance for a correct design and implementation. In one project where we were hired in as an independent quality consultant, the requirement specification was a major problem throughout the project (Skogstad & al., 1991). The requirement specification was neither complete nor consistent. It included reliability requirements that was impossible to test and it underwent several changes throughout the project due to increasing understanding of the problem. During all phases of the project we found that most errors were caused by the unstable requirement specification.

Still the same question is valid for every new project: where are the pitfalls in this particular project?

The Avoidance, Detection and Removal of Faults

Traditionally, validation of software is tantamount to removal of faults during in the testing phase. Lipow (1977) indicates that the number of faults left after testing has a linear relationship (1:10) with the faults found during testing, i.e. the more faults that are found, the more faults are probably left in the system. Common experience is the same: the more faults found during testing, the more unreliable the software component will be during operation.

During the last years fault avoidance techniques have got increasing attention. Testing alone does not give us confidence that the reliability goals are reached. The fault avoidance techniques must be an essential part of the quality assurance activities. Both fault avoidance and fault detection techniques should be used together in order to give the user confidence that the reliability goals are met.

The Fault Tree technique is a fault avoidance technique in that it can be used to identify parts of the software developing process that need special attention. The quality assurance techniques include avoidance as well as detection techniques since it contain reviews and testing as well as recommendations for good programmers practice.

USING FTA TO IDENTIFY FAULT INTRODUCTION

The Fault Tree Analysis Technique

The Fault Tree technique is a commonly used technique of risk and reliability studies. A Fault Tree is a logic diagram that displays the interrelationships between a potential critical event (accident) in a system and the causes for this event. The examples in the paper are constructed by the help of a computerized Fault Tree tool developed by Sintef.

The Fault Tree technique is here used on a software development process. An example is used to show an application of the technique. The goal is first to find critical events concerning fault introduction. Secondly, it is to relate those events to the developing process in order to find appropriate actions that will reduce the probabilities of that introduction.

The overall problem with the developing of software, or any system, is human imperfection. In all branches of the Fault Tree there should have been an event that say "the fault is due to human imperfection". However, since this will give us a model of the process that leave us with unsolved problems and no hope of progress, we will skip this event on the top levels.

An Example

The information necessary to build the Fault Tree can be obtained from the following sources:

* project plans, requirement specifications, pre-project results

* experience from earlier projects done by the same company

* experience from relevant projects in the literature

* information obtained during the developing process. The more we know about design, implementation and quality assurance results, the more precise our estimates for the top event will be.

Fig.2 is an example on the use of Fault Tree technique on a software development process. Our goal is to identify the basic events that will cause the top event to occur.

The most important event as seen from the developer is "Failure after release". This event is thus chosen as the top event in our Fault Tree. The Fault Tree in fig.2 can be described as follows:

* the top event will occur if a failure occurs in one of the modules that constitutes the system

* the failure of one of the modules will occur if there is a fault in the module AND the module is activated during a run

* the fault in the module is a remaining fault from one of the phases of the developing process

* faults are carried through from one phase to the next because the developer initially made an error AND because the fault was not discovered by the quality assurance (QA) activities of that or any later phase.

The bottom events of fig.2 can be developed further by asking the following questions:

* why are errors done in this particular phase?

* why are the faults not discovered by the quality assurance activities of that phase?

In fig.3 the event "faults not discovered by QA activities" is further developed. The basic events in this example are, according to our experience, common events in most projects:

* quality assurance goals are missing or imprecise

* quality assurance methods are not defined or incompletely implemented. Our experience is that this is probably the event responsible for most of the unsuccessful quality assurance activities

* the quality assurance plan is missing or incomplete

* deviation from standards are tolerated or not even noticed. This can happen if the developing company has no overall policy for developing software.

* sloppiness, which is a kind of human imperfection all too often present

In fig.4 the event "an error is made in phase y" is further developed. The basic events in the example are one of the following:

* specifications are incomplete, i.e. they are either missing, incorrect or misinterpreted

* misinterpretation of specification may be caused by lack of knowledge on application AND lack of communication between user and developer

* incorrect use of tools that may lead to errors

In a real project all these events have to be further developed in order to find the possible reasons for the error done and as a consequence, actions that can solve the problem. E.g.: lack of knowledge on external environments may lead to misinterpretations as well as incorrect use of tools, i.e. the problem to be solved is lack of knowledge.

The Analysis

The goal of the analysis is first to identify the basic events that are responsible for most of the fault introduction. Secondly, we need to make a list of those events that can be manipulated in order to increase the probability of success of the developing process, i.e. reduce the probability of the top event.

The analysis can be done in two ways:

- a qualitative approach. This is done by ranging the basic events according to their likelihood. A possible way to do this is to use the terms ´not likely´, ´possible and ´very likely´. The top event can then be given a qualitative assessed on the basis of the assessment of the likelihood of each basic event. By using a qualitative approach, it is possible to bypass a long and tedious discussion of probabilities. In a qualitative approach the preceding event will

  - at an AND gate take the "lowest" value of the input events

  - at an OR gate take the "highest" value of the input events

- a quantitative approach. In this approach we must assign numerical values to the basic event probabilities. The top event probability can then be found in the standard way. Sources of information on the basic event probabilities may be found in the literature and in our case, from our own experience. However, it is not obvious how to estimate these probabilities. Software is not worn out like mechanical systems and given guarantees on MTTF or MTBF. When dealing with software it may be expedient to compute a worst and a best case concerning the top event probability.

Identification of Important Phases

Our goal is to find activities that can be manipulated in order to avoid fault introduction. From the Fault Tree we can extract the most critical basic events. The next step is obvious, but not simple: we have to introduce activities in the appropriate developing phase, so that the probabilities of the basic events in question are reduced. It is important that the Fault Tree analysis is done before it is too late for corrective actions. Retrospective analysis may be good for the collecting of experience, but it is of no use for the project that is analysed.

E.g.: to reduce the probability of the event "misinterpretation of specification" we have to

- train the developer before commencing with the activity

- determine and / or develop the appropriate standards to be used

- introduce one or more walk-throughs during and at the end of the specification phase

All these activities require resources that has to be planned for in the budget and in the project plan. The identification of project phases that has to be replanned is the practical result of the analysis. If this is not done, the probability of success will suffer.

The Importance of the Construction Process

The insight gained during the Fault Tree construction is just as important as the information we can extract from the Fault Tree when it is finished. Our experience is that even those who have never seen a Fault Tree before, grasp the idea after a short introduction and are able to discuss the basis for the analysis and come forth with their own ideas. A structured way of analysing a process helps the users to see the problems from a new angle and thus find new solutions.

It forces project management and developers to take a stand concerning project goals, resources allocated to different phases and the use of quality assurance activities during the developing process

A Fault Tree is easy to change if we have a tool. It is then possible to use the Fault Tree during the whole process, to add events and change probabilities as the developing progresses and new insight is gained. The Fault Tree can be used to show the consequences of the changes, answering the important "what if.." questions.

Instead of constructing a complete Fault Tree of the whole developing process, it may be more practical to focus on the most probable and relevant parts. In this way, we avoid tedious discussions on events and relationships that do not contribute significantly to the final results. Later in the process, the Fault Tree can be elaborated on when new events occur.

USING QA TO REDUCE FAULT CONTENTS

Quality assurance

The goal for the quality assurance is to ensure that quality is being built-in as the system develops, rather than being discovered at the end of the process. The quality assurance activities should be planned at the start of the process and be an integrated part of the project plan. Different quality assurance techniques are in common use: walk-throughs, inspections, code reading, and audits as well as the traditional testing.

Co-Working events

We will now show how the Fault Tree can be used when we try to change the process by introducing quality assurance activities.

A Fault Tree for a reliable system should have an AND gate at the top event. An OR-gate at the top event tell us that the probability of the top event may be high. In our example in figs. 2-4 there is an OR gate at the top event and just a few AND gates are present in the whole tree.

Let us return to our example. Our experience tell us that the following events are "very likely" to happen

- an incomplete specification is developed
- quality goals are not met
- deviations from standards are tolerated or not noticed
- quality assurance methods are incompletely implemented
- functional specifications are misinterpreted

It is easy to show how these events combine to cause the top event. The Fault tree can in this case be used as a map where the possible routes to failure are marked out.

Notice the difference between the AND and OR gates. The events "standards are not followed" and "deviation from standards are tolerated or not noticed" can be stopped at an AND gate. At OR gates the faults pass through without any problem.

One approach to reduce the probability of-the top event is to try to change the process so that we get more AND gates in the Fault Tree, i.e. introduce events that stop the propagation of faults.

Fig.5 shows how introduction of AND gates can be done in a practical way. The event "deviation from standards are tolerated or not noticed" is very likely to happen. Quite often we do not notice deviations from agreed procedures or we tolerate it due to lack of time, lack of resources and because we believe it won`t matter much.

The following combination of events will reduce the probability of deviations from and negligence of standards:

- the developing of an overall company policy on quality assurance
- the use of technical walk-throughs and a strict follow-up system
- demands from the customer on quality assurance activities

The human factor is introduced in the event "lack of understanding of the importance of the quality assurance activities". From this diagram we see that it is very important that the developer

- understand why quality assurance must be carried out
- know the effects of quality assurance activities
- know how to perform the various quality assurance activities.

With the negligence of these factors and a lousy follow-up system, the top event will occur with a high probability.

As a conclusion on this we can say that it is essential for a reliable development system that all through the developing process

- every activity has a corresponding quality assurance activity
- both the event AND the quality assurance activity must fail for the next level event to become true.

It these conditions are fulfilled, the probability of the top event will decrease and as a consequence, the probability of success increase.

CONCLUSION

Fault Tree technique is well suited to analyse the software development process for several reasons:

- the uncertain factors in the production process and their combinations are made visible. The graphic representation made possible by the Fault Tree is easy to communicate both to managerial and technical personnel
- it is easy to show the consequences of changes in the process, thus answering "what if.." questions
- it helps us to focus our attention on the most critical parts of the process and thus pointing out possible answers to how and when faults are introduced into the software

REFERENCES

Basili, V.R. and B.T. Perricone (1984). Software errors and complexity: an empirical investigation. *Communications of the ACM*, Vol. 27, no. 1

IEEE Standard Glossary of Software Engineering Terminology, IEEE Std 729-1983, IEEE Inc., 1983

Laprie, J.C. (1989). Dependability: a unifying concept for reliable computing and fault tolerance. In T. Anderson (Ed.), *Dependability of resilient computers*. BSP Professional Books, Oxford. pp.2.

Lipow, M. and T.A.Thayer (1977). Prediction of software failures. *Proceedings 1977 annual reliability and maintainability symposium*. pp.489-494.

Sayet, C. and E. Pilaud (1990). An experience of a critical software development. *IEEE Conference Proceedings. Fault tolerante computing: 20th International symposium*. pp.36-45

Skogstad, Ø., E.Ø. Øvstedal and T. Stålhane (1991). Applying quality assurance in a project with conflicting interests. In P. Ancilotti and M. Fusani (Ed.), *Proceedings from the international conference on achieving quality in software*. ETS EDITRICE, Pisa Italy. pp.473-496.

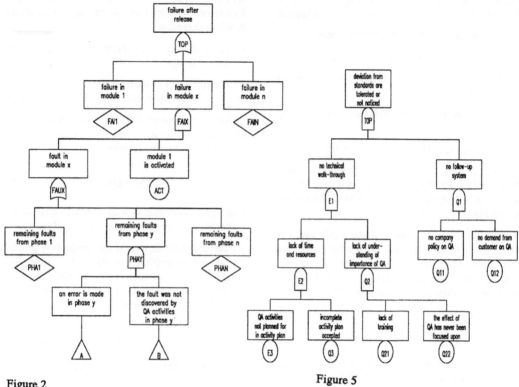

Figure 2

Figure 5

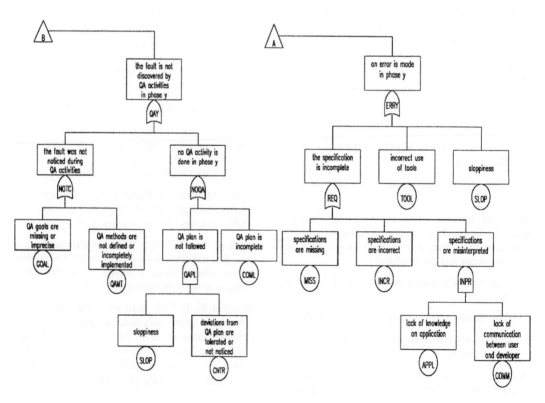

Figure 3

Figure 4

SYNCHRONIZED LOCAL STATE DIAGRAM: A MODELLING METHODOLOGY UNDER DEVELOPMENT[+]

P. E. Heegaard*, B. E. Helvik*/** and E. Gotaas**

*SINTEF DELAB, N-7034 Trondheim, Norway
**Norwegian Institute of Technology, N-7034 Trondheim, Norway

Abstract: A sufficiently truthful dependability modelling and analysis of large distributed telecommunication systems poses a number of problems, not yet resolved.

To overcome the weaker sides with current modelling methodologies, a new modelling methodology is suggested in this paper called Synchronized Local State Diagrams, or for short SLSD. The basics of SLSD are:

- a unit or a group of units are described by a local state diagram,

- dependencies are modelled by transition rates which are functions of the global state of the system, and by synchronization primitives,

- intact/defect criteria of the system are defined by fault trees which have states of the local diagrams as their leaves.

SLSD uses an object oriented technique for building models of large systems.

Keyword: Modelling, dependability, object oriented technique, local state diagrams, Markov processes.

1 Introduction

The objective of this paper is to introduce a modelling methodology denoted Synchronized Local State Diagrams, SLSDs [1]. SLSDs are aimed at dependability evaluation of large telecommunication systems, systems with hundreds of processing elements of various types, many of them controlling peripheral units, an interconnecting network, an extensive maintenance and operation subsystem, a large and complex software, etc.

In addition to be able to handle large systems, to be easy to read and understand, the following properties of the system must be reflected in the model:

The dynamics: This aspect concerns both the system behaviour, e.g. how the specific sequence of events influencing the dependability, and its dependability measures, like the distribution of the time between failures, see [2].

Dependent units and subsystems: Dependencies between the elements in a system are often neglected in dependability modelling and analysis. The consequence of this is usually too optimistic predictions. Three classes of dependencies are considered:

- *Error propagation*, e.g. an inconsistency in one processing unit causing an error in another.

- *Common resources*, e.g. a set of units share a limited number of repair-men.

- *Imperfect fault/error handling*, deficiencies in detection, location or error isolation may put more than the erroneous node out of service, e.g. imperfect coverage [3].

Why develop a new modelling methodology instead of using the currently available techniques? Cf. tutorial papers [4, 5] for thorough discussions of these. Unfortunately, they all have drawbacks with respect to the above outlined objectives. *Block diagrams* and *fault trees* enables modelling of large systems and are easy to read and understand. They are unable to model the dynamics of the system and are restricted to independent units. Both dependencies and dynamics are well handled by *global state diagrams* and *Petri nets*. The latter has its strength in modelling synchronism and parallelism, but it has a high application barrier. Neither state diagrams nor Petri nets are suited for modelling of large system due to explosions of the number of states/places, which makes them difficult to read/develop for other than small systems. A modelling technique should, in addition to be able to include the dynamics and the dependencies of a system, also be user friendly.

Local state diagrams is still under development and its pro and cons are not yet fully explored. The rationale for introducing them is to avoid the state space explosion of the global state diagrams, and at the same time maintain the advantages of state diagrams like simplicity, ease of understanding and ability to model dynamic behaviour.

Hence, a modelling methodology, SLSD, which intend to meet the above outlined requirements, is proposed. It is based on the following basic ideas, picking the best properties from both global state diagrams and Petri nets and avoiding the drawbacks.

- A single unit or a group of units are described by a local state diagram. A unit may either be a structural or functional part of the system.

[+] The work reported in this paper are done under contracts with Norwegian Telecom, Research Department.

- Dependencies are modelled through linking local state diagrams by transition rates which are functions of the global state (states of the other local diagrams).

- Global events, which is reflected by a simultaneous transition in more than one local diagram, is handled by a synchronization (Petri net like) primitive.

- Intact/defect criteria of the system with respect to various dependability levels, are defined by fault trees. The leaves of these trees are formed by states of the local diagrams.

The next section describes the local diagrams, and Section 3 outlines their use for building system dependability models. In the last section some concluding remarks are given.

2 Local diagrams and their mechanisms

This section contains a description of the notation and mechanisms of the local state diagrams in SLSD. A *state* is a set of characteristics, describing the interesting properties of the entire system (*global* state) or a subsystem/unit (*local* states). The global state is given by combinations of local states. Between these states, *transitions* may take place. The rate of transitions are constant and depends solely on the (global) state. The system are therefore described by a discrete space continuous time Markov process, see for instance [6]. Non homogeneous models (e.g. to model reliability growth) and semi Markov models (non n.e.d. state sojourn times) are not considered so far.

In the proposed method, subsystems/units are modelled separately irrespective of whether they are independent or not. SLSD allows both a structural and a functional decomposition of the system. A separately modelled part is referred to as a *component*. Allowing dependences between these components introduces linkage, as described in the following subsections. In addition, explicit modelling of contention on shared resources may be useful. This is not yet fully investigated.

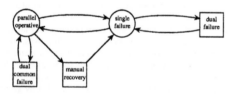

Figure 1: Example of a local state diagram, modelling a parallel synchronous subsystem with imperfect coverage.

In Figure 1 a local state diagram consisting of states and transitions are given, modelling a parallel synchronous subsystem with imperfect coverage, see [3].

2.1 Transition rates

The rates assigned to the transitions between states in a local diagram are not necessarily constant, but can be expressed as a function of the states of other diagrams, i.e. the global state. This models effects like:

- *error propagation*,

- *repair dependencies*,

- *additional stress/load introduced rate changes*,

- *state dependent failure rates*.

The general transition rate may be expressed as a function of the global state space Ω to include all effects above. The rate is partitioned into a sum of contributions of various *local* state space variables, ω_l. In the current approach, this is restricted to a linear combination, expressed as:

$$\theta_{ij}^{(k)}(\Omega) = \lambda_{ij}^{(k)} + \sum_{l \in [1..N]} [f_l(\omega_l)]_{ij}^{(k)} \qquad (1)$$

where $\theta_{ij}^{(k)}(\Omega)$ expresses the transition rate from state i to j for the local diagram k, $\lambda_{ij}^{(k)}$ is the basic transition rate, and $[f_l(\omega_l)]_{ij}^{(k)}$ is the (non negative) contribution on transition $i \rightarrow j$ in diagram k from diagram l as a function of its local state ω_l; $l = 1, \cdots, N$ where N is the number of diagrams, and $\Omega = \{\omega_l\}_{l=1,\cdots,N}$, this is similar to the one introduced in [7].

2.2 Global event synchronization

A mechanism which handles *global events* must be included. A global event is an event that causes state changes (transitions) in more than one separately modelled unit, subsystem or function, simultaneously. A global event is constituted by a set of local events. Local events are reflected by *mandatory* and *optional* transitions in corresponding diagrams. Below, an example on local and global events are given.

Example 2.1 *Given a set of units, which is either passive or active. They share a resource which is either available or unavailable. When one unit goes active, the resource is granted and it becomes unavailable if it was available at request. The unit activation and resource occupancy are local events, though they must occur simultaneously. The corresponding global event is therefore constituted by both activation of a unit and a simultaneously allocation of the resource. Note that while the resource is occupied, no resource requests from the passivated units may occur.*□

For a global event to take place, the conditions for that event must be fulfilled, e.g. from the example above the resource must be available. A condition corresponds to that the local diagrams which have mandatory transitions are in their originating states for those transitions. A mandatory transition will always take place if the global event occur. An optional local transition will take place if the corresponding diagram is in its originating state when the global event occur (the slave transition is an artifact which allows us to avoid a number of nearly identical global events).

Among the transitions causing the events constituting the global event, we have one *master* and one or more *slave* transitions. The conditions for the global event and its rate is associated with the master. The choice of a master transition must be done among the mandatory transitions and should be done according to the properties to be studied. However, a transition can not be master transition for one global event and slave transition for another.

Figure 2 shows the symbols introduced for modelling of the synchronization mechanism. Arrows pointing *at* the triangles indicates master transitions (*sending* a signal), and arrows in the opposite direction indicates slaves (*receiving* a signal triggering the transition). Solid and dashed triangles indicates mandatory and optional slave transitions respectively.

Inside every synchronization symbol there are one or more references. Master transitions uses it both as a reference and as condition for the mandatory slaves, and only as reference for optionals. The slave transition uses the reference to identify its master transitions.

The reference notation used both to and from the triangles is hierarchical constructed to make it easier to find the relations:

<diagram type> : < diagram no>.<state no (to state no)>

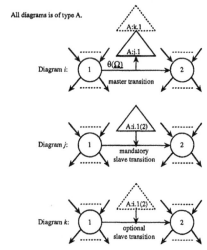

Figure 2: Special synchronization symbols.

The rational for the diagram type, diagram no. convention is given in the next chapter. The condition is given as a logical function of the references. For instance if references/conditions are ANDed, several mandatory slave transitions exist for that specific global event. If the transition includes an OR-list of conditions, this means that the transition in question is master transition in more than one global event.

3 Building system models

When large systems are modelled, the number of diagrams will become very large. Many of the diagrams will be quite similar, and differ only in their identity and perhaps the transition rates. Bookkeeping problems will also arise in keeping track of all the diagrams constituting a large model and in keeping them consistent. In this chapter the principles behind building models of large systems are presented. The idea of building large systems from previously defined diagram types, modelling types of units and subsystems, are adopted from object oriented programming [8].

3.1 Diagram types and components

As pointed out in the introduction, it is foreseen that both system units, subsystems (formed by many units), and functional elements, may be modelled by one separate local diagram type. An element, in this context, is for instance a processing unit with both hardware and software. Elements may also be strictly functional. For instance, it has been fruitful to model the "state of a common media protocol by a separate diagram.

It may be advantageous to model more than one element by one diagram. For instance, a group of identical processing elements performing the same task may conveniently be modelled by one diagram type. The states of this local diagram will be defined by the number of processing elements in their various operational modes. This will result in fewer global states. It is also natural to model very closely coupled units by one diagram, e.g. a pair of processing elements working in parallel synchronism, see Figure 1. In a system there may be a number of elements and/or sub-

systems which are similar. The concept of an Element-Diagram is therefore introduced.

Element-Diagram is a local state diagram of a type of system element/unit or a type of subsystem, where all the states and transitions between the states are defined. Only transitions which have the same transition rate in all occurrences of that type are assigned this rate. An Element-Diagram is represented by a circle in the system graph that will be introduced in Subsection 3.2.

In the same way as larger systems are made up by subsystems, which again are made up by subsubsystems, the modelling approach includes grouping of Element-Diagrams into System-Components in an arbitrary number of levels. It is not desirable to lock the model into a strict hierarchical scheme. Hence, the following flexible definition.

System-Component is formed by none or more Element-Diagram(s) and none or more (sub)System-Components. It must be formed by at least one Element-Diagram or by one (sub)System-Component and another (sub)System-Component/Element-Diagram in order to be meaningful. Transition rates, inclusive the functional relationship, and the synchronization between its Element-Diagrams and (sub)System-Components are defined. All the rates need not to be defined, but they may be. The System-Component represents either a type of subsystem or a specific subsystem. If all the rates are defined, the subsystem is independent of the others. The System-Components are represented by rectangles in the system graph introduced in Subsection 3.2. A fault tree may be associated with the System-Component. The fault tree represents the dependability structure of the component. This tree may have trees of its (sub)System-Components as branches. See Subsection 3.3.

System-Model is the root System-Component. All rates, which is not defined in the underlying System-Components, must be defined in the System-Model. Similar to the ordinary System-Components, the System-Model is represented by a rectangle in the system graph introduced in Subsection 3.2.

3.2 The system graph

The system graph is a directed acyclic graph used to define how the System-Model is built from the Element-Diagrams and System-Components related as outlined above. The root node is the System-Model and the Element-Diagrams defined for the model are the leaf nodes. The arcs go from each System-Component to its Element-Diagram(s) and/or (sub)-System-Components. A number on the arc gives the number of Element-Diagram(s) or (sub)System-Components of this type the System-Component is formed by. Keep in mind that the relations between (sub)System-components are defined in the current System-Component. Hence, what is "seen in a System-Component are all the local state diagrams which defines this component. The functional relation between the rates of all these diagrams and the synchronization between them are defined. For clarification, regard the example depicted in Figure 3.

Example 3.1 *First, regard System-Component Single_U. It is made up by two different local diagrams, proc_e and periph. The proc_e models a general processing element and periph the equipment it controls. The dependencies between these two system elements are represented in Single_U. Proc is a processing element without peripheral equipment. Double_U is a pair of subsystem as represented in Single_U and the interworking between these is modelled by interw.*□

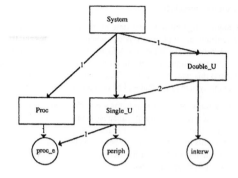

Figure 3: System graph defining eight local diagrams.

The system graph shows the different types of local state diagrams which describe the system. By the system graph inclusive the definitions in the System-Components, the various local diagrams of the system are completely defined. These individual local state diagrams are generated by traversing all possible paths through the graph. Counting the paths through the system graph in Figure 3 shows us that this system is modelled by eight local diagrams of three different types.

3.3 Defining dependability levels

In large and complex systems like telecommunication switching systems, there are not a single intact – defect criteria. Several dependability levels exist, where each level is associated with a service degradation in terms of number of affected subscribers, affected traffic volume, performance reduction, etc [2]. Each of these dependability levels may be regarded separately.

A local defect state can be, but is usually not, a global defect state. Note also that a global defect state (with respect to a dependability level) need not to be represented by a local defect state or a combination of such states. It is therefore necessary to have criteria which classifies every global state as an intact or defect state with respect to the various dependability levels. A global state is uniquely defined by a combination of local states (as long as no non-Markovian scheduling is applied, like FIFO and pre-emptive priorities).

It is suggested to model the dependability structure of a system, with respect to the various dependability levels, by *fault trees*. Specific for these trees is that *the leaves are states in their local diagrams*. Hence, the trees will define the combinations of local states which represent system defect states.

Trees may be, but need not to be, defined for all the System-Components. A tree may use trees of (sub)System-Components as branches. A fault tree must be defined for the System-Model. The top event in the System-Model tree is that the system cannot meet the intact criteria of the actual dependability level. The approach is illustrated by following example.

Example 3.2 *The system regarded has three elements (subsystems). These are of two types as sketched in Figure 4. Note that* SlsdB *has three states – in addition to the* OK *and* Down *states a* Degraded *state where the subsystems of this type have a lower performance.*

In Figure 5(a) the fault tree associated with Comp_B_1 *is shown. It reflects the same as the square state symbol in Figure 4(a). The fault trees associated with the other system components are similar. Figure 5(b) shows the fault tree of the entire system. It is seen that the system has failed if one of the system components has failed or if both subsystems of type B are in their* Degraded *state.*□

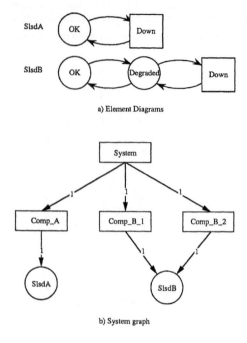

a) Element Diagrams

b) System graph

Figure 4: Models of the system in example 4.3.

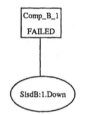

a) Fault tree assosiated with system component B_1.

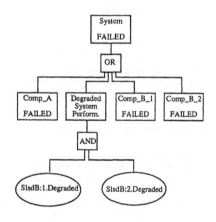

b) System Fault tree.

Figure 5: Fault trees defining the defect states of the system in example 4.3.

It is left for further study to investigate how these trees may be arranged for the most efficient modelling, both hierarchical towards higher dependability levels (more serious degradations) and hierarchical within the system structure as modelled by the system graph.

4 Concluding remarks

A new modelling methodology for building large system models is introduced. It is called Synchronized Local State Diagrams (SLSD) and is based on local state diagrams modelling the basic functional or structural elements of the system. Addition of new mechanisms enables building of more refined models, taking into account various types of dependencies between the elements.

SLSD allows elements to be treated separately, in an easily readable and understandable way, without oversimplification, state explosion and unrealistic independence assumptions.

A prototype tool supports the methodology by:

- A description language for input of system models

- Regenerative and block simulation of the model estimating the stationary unavailability.

- Automatic generation of *full* (global) transition matrix from the SLSD description.

is available.

The methodology is under development and some work still remains. The most important outstanding issues are:

1. Further investigations on how resource contention (scheduling procedures, priorities, etc.) can be incorporated. This in order to facilitate the modelling of repair strategies.

2. Inclusion of accelerated simulation based on importance sampling, using SLSD model information, to speed up evaluation or improve the accuracy for highly reliable systems.

References

[1] Poul E. Heegaard and Bjarne E. Helvik. Synchronized local state diagrams; description of a modelling method under developement. Technical Report STF40 A90100, ELAB-RUINT, May 1990.

[2] Poul E. Heegaard and Bjarne E. Helvik. Short term measures, dependability-levels and alarm-levels as means for dependability for dependability evaluation of switching systems. Technical Report STF40 A90099, ELAB-RUINT, May 1990. In Norwegian.

[3] T. F. Arnold. The concept of coverage and its effect on the reliability of a repairable system. In *FTCS-2 - The second international symposium on fault-tolerant computing*, 1972.

[4] Allen M. Johnson jr. and Miroslaw Malek. Survey of Software Tools for Evaluating Reliability, Availability, and Serviceability. *ACM Computing Surveys*, 20(no. 4):227–269, Dec. 1988.

[5] M. Mulazzani and K. Trivedi. Dependability Prediction: Comparison of Tools and Techniques. In *Proc. IFAC SAFECOMP'86, Sarlat, France*, pages 171–178, Oct 1986.

[6] D. R. Cox and H. D. Miller. *The Theory of Stochastic Processes*. Chapman and Hall, 1965.

[7] Bjarne E. Helvik. Modeling the Influence of Unreliable Software in Distributed Computer Systems. In *FTCS-18 - The eighteenth international symposium on fault-tolerant computing*, pages 136–141, June 1988.

[8] Eigil Gotaas. Generation of a state-space representation from system structure. Master's thesis, The University of Trondheim, The Norwegian Institute of Technology, May 1990. In Norwegian.

BALANCING RELIABILITY REQUIREMENTS FOR FIELD DEVICES AND CONTROL LOGIC MODULES IN SAFETY SYSTEMS

L. Bodsberg and P. Hokstad

SINTEF Safety and Reliability, N-7034 Trondheim, Norway

Abstract. The paper provides guidelines for the definition of reliability requirements for computerized safety shutdown systems in the process industries. The main question discussed is how to derive safety system requirements which ensure that the reliability of field devices and control logic modules is balanced from a safety point of view.

Reliability figures of example safety systems are presented for various configurations of sensors, input/output cards, Central Processing Units (CPUs), and actuating elements. The figures are based on quantitative reliability analyses using a model and methodology for probabilistic safety assessment developed by SINTEF. The main new feature of this model compared to other models is that the effect of all types of failures occurring during field operation is considered in an integrated manner. Thus, failures due to excessive environmental stresses and human-made mistakes during engineering and operation are included in addition to failures due to natural aging of components (inherent failures). The effect of self-test is also included in the model.

Keywords. Reliability; safety; instrumentation; optimal systems; industrial control.

INTRODUCTION

The main components of a computer-based safety system are:

- Sensors (e.g. gas detectors, pressure switches)
- Control logic units (input/output cards, Central Processing Units (CPUs)
- Actuators (e.g. valves, fire dampers)

Sensors and actuators are termed field devices. The sensors detect abnormal operating conditions and provide signals to a logic unit. This unit performs logic manipulation (voting) of the signals and generates output signals which automatically activate appropriate safety actuators. Operators receive alarms and monitor system performance via displays in the central control room.

The design process of a safety system should start with a set of basic system requirements; including requirements for system effectiveness and life-cycle cost (LCC). The life-cycle cost is the total cost for the user to purchase, install, use, and maintain the system. The overall objective is to design safety systems which fulfil stated requirements for safety performance, and which at the same time minimize LCC.

Both the user/purchaser and the vendor of safety systems may take advantage of quantitative reliability and life cycle cost analyses to identify system configurations and operating philosophies which are the most cost effective regarding safety.

A reliability and life cycle cost prediction model termed the PDS model (Aarø, Bodsberg, Hokstad, 1989; Lydersen, Aarø, 1989; Hokstad, Bodsberg, 1989) has been developed by SINTEF to assist the designer in identifying the preferred system configurations and operating philosophies. The main new feature of this model compared to other models (Poucet, Amendola, Cacciabue, 1987) is that the effect of all types of failures occurring during field operation is considered in an integrated manner to obtain results which reflect *operational* rather than *theoretical* system performance. Thus, failures due to excessive environmental stresses and human-made mistakes during engineering and operation are

included in addition to failures due to natural aging of components (inherent failures). The effect of self-test is included in the model.

The main question of this paper is how to derive safety system requirements which ensure that the reliability of field devices and control logic modules is balanced from a safety point of view. In a balanced system the contributions to overall reliability from field devices and from logic control modules are in the same order of magnitude. Main steps in the definition of system requirements specifications are proposed. Further, reliability figures of safety systems are presented for various configurations of sensors, input/output cards, Central Processing Units (CPUs), and actuators. The figures have been established using the PDS model. A description of main features of the PDS model is included in the paper.

Input data to the quantifications are based on field data provided by oil companies and vendors of control and safety systems. Field data have been supplemented by judgement of company experts.

Results concerning LCC are not included in this paper, but it is stressed that safety measures must be evaluated with respect to cost-effectiveness.

SYSTEM SPECIFICATIONS

When establishing specifications for safety systems, the total function of the system should be addressed. Functional requirements for the total system, including the detectors, the logic control units, the actuating elements (e.g. valves), should be stated.

The proposed steps in specifying the system requirements are:

1) **Identify the task of the safety system and the corresponding safety criticality**

   The safety requirements for the system should be related to the tasks of the system, e.g. to shut in the production wells upon fire in a fire area. The user of the safety system should identify these tasks and the safety criticality of the tasks.

2) State the overall safety requirements

The minimum acceptable safety performance should be given as quantitative safety requirements (e.g Critical Safety Unavailability - defined later) for each of the tasks identified by the user.

3) State the safety requirements of each subsystem

Overall safety requirements should be allocated to the various subsystems. The subsystems of the safety system are the detector part, the logic control units, and the actuating elements, e.g. valves.

This means that one uses a top-down approach when stating specifications for the safety system, resulting in a hierarchical structure of the specifications.

When establishing safety requirements for the detectors, the logic control units and the actuating elements, extra care should be taken to ensure that the requirements are balanced.

PDS MODEL

In the PDS model, the effect of component *suitability/capability* is modelled. Given that a component will operate as *designed* when called upon, does not necessarily ensure that the component will be able to perform its specified service from an operational point of view. For instance, the specified service of a fire detector is to detect the presence of a fire. However, fire detectors detect only fire *phenomenon* like heat, smoke, flame, and one particular type of detector is generally suitable for detecting only one type of fire phenomenon. Installing detectors which have high technical reliability is of little help if the detectors are not suitable for the type of fire conditions occurring in the area. (Flame detectors are unsuitable for detecting fire conditions with a lot of smoke.) These considerations, which seem fairly obvious, are seldom considered in RAM analyses.

QUANTIFICATION OF SAFETY PERFORMANCE

The following measure is used to quantify safety performance in the PDS model:

■ **Critical Safety Unavailability (CSU):**
The probability that the safety system will fail to automatically carry out a successful safety action on the occurrence of an abnormal operating condition during a period of time in which nobody is aware that the safety system is unavailable.

Some comments on the elements in this definition are given below:

The term *probability* represents a fraction, specifying the number of times that one can expect the event to occur in a total number of trials. For instance, a CSU equal to .01 indicates that the safety system will fail to carry out a successful safety action in 1 out of 100 occurrences of an abnormal operating condition.

The term *automatically* indicates that manual activation of the safety system is not included. Thus, CSU does not reflect the probability that operators activate the safety system manually.

The term *successful* indicates that a specific success criteria must be established which describe what is considered to be successful. It should be noted that the term successful relates to the function (service) the safety system is to deliver. Thus, CSU is always related to one specific type of abnormal condition.

A precise success criterium for one particular function is: "The safety system shuts in all DHSVs (Down Hole Safety Valves) upon

hydrocarbon fire in Fire Area 1". The corresponding undesirable event is: "Failure of the safety system to shut in all DHSVs upon hydrocarbon fire in Fire Area 1".

The term *nobody is aware that the safety system is unavailable* indicates that physical degradation of components is not considered to be critical from a safety point of view after it is revealed by built-in or manual tests. Nor is unavailability of the safety system during testing or repair (system in by-pass mode) included in the CSU figure. During testing and repair it will be known that the safety system is unavailable. This situation will not necessarily represent a safety hazard, but rather requires some additional precautions. For instance, if fire detectors are bypassed, personnel located in the area may be tasked to activate proper shutdown actions manually. If this situation represents a hazard, the production should actually be shut in. Failure of emergency shutdown valves often requires that production is shut in.

FAILURE CLASSIFICATION SCHEME

When using the PDS model, the component failure rates/probabilities are specified with respect to failure mode and failure cause.

Component failures for safety systems are split into the following three failure modes:

■ Fail-To-Operate (FTO) upon demand
■ Spurious Operation (SO)
■ Non-critical (NC), failure will neither cause loss of safety nor spurious trips

The first two of these failure modes, Fail-To-Operate (FTO) and Spurious Operation (SO) are considered most important, as they might lead either to loss of safety or production unavailability cost.

Figure 1 shows the classification scheme for failure cause which is used in the PDS model.

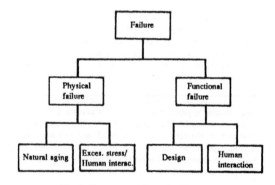

Fig. 1. Component failure classification.

■ **Physical failures**, the delivered service deviates from the specified service due to physical degradation of the component.

Any failure which requires some kind of *replacement* to restore the component service may be denoted physical failure. All failures caused by natural aging and environmental stresses belong to this category.

■ **Functional failures**, the delivered service deviates from the specified service although there is *no physical degradation* of the component.

Replacing the component with an identical component will not remove the failure cause. Modifications rather than repair or replacement actions are required in order to remove the failure cause. Functional failures may be grouped into design failure and interaction failure.

The two categories of physical and functional failures are more fully described in the following:

■ Physical failure due to *natural aging* occurs under conditions within the design envelope of a component.

■ Physical failure due to *stress/human interaction* occurs when excessive stresses is placed on the component or when personnel make mistakes during operation.

■ Functional failure due to *design* is initiated during engineering and constructions (and is latent from the first day of operation).

   This failure will not be identified by periodic testing of the system during operation. If the failure is not detected during the very first operational test, it will not be detected by subsequent tests either. Thus, this failure is also denoted *Test-Independent Failure* (TIF). Redesign is required to remove failure cause. If the failure cause is removed and no other failures are introduced, higher reliability will be achieved.

■ Functional failure due to *human interaction* is initiated by human made mistakes during operation. This failure may be detected by periodic testing.

The natural aging failures are *independent* failures. All other failures are *dependent* failures, as they might involve the failure of more than one component, when hardware redundancy is applied.

BALANCED CONFIGURATION - FIELD DEVICES VS. CONTROL LOGIC UNITS

Figure 2 shows the predicted safety performance in terms of critical safety unavailability of various voting logics of safety system modules. The letter *s* is used for CPU configurations with self-test. The voting logics are here defined as follows:

- 1oo1 (One-out-of-one) There is a single module. This module has to give a shutdown signal for a shutdown to be activated (no redundancy - no voting).

- 1oo2 (One-out-of-two) There are two modules. Only one of these has to give a shutdown signal for a shutdown to be activated.

- 2oo2 (Two-out-of-two) Both modules have to give a shutdown signal for a shutdown to be activated.

- 2oo3 (Two-ot-of-three) At least two of the three modules have to give a shutdown signal for a shutdown to be activated.

According to Fig. 2, the safety performance of input/output cards is generally better than the performance of detectors, CPUs and valves. For instance, a single input card has lower CSU than a duplex CPU unit with 1oo2 voting and self-test. Thus, *single* input/output cards and *redundant* sensors/actuators are generally an optimal configuration from a safety point of view. Hardware redundancy of input cards gives only a moderate overall improvement in safety performance.

System vulnerability must be addressed in addition to reliability when determining the hardware redundancy/voting logic of CPUs. The fact that the CPU is demanded by all types of abnormal operating events (gas leak, fire and overpressure) and that a failure of

the CPU may prevent any actuator to carry out intended safety action, often implies that the CPU should be duplicated. Two CPUs with 1oo2 voting and self-test is optimal when there is low/medium cost per trip. Two CPUs with one CPU in standby is also a feasible option. Three CPUs with 2oo3 voting and self-test should be considered when there are very high costs per trip.

Valves contribute significantly more than output cards to CSU. However, the contribution from the valves to system CSU is negligible when three different safety valves are used to shut in a well.

Figure 2 shows that 1oo2 voting logic is the best configuration whereas 2oo2 is the worst configuration from a safety point of view. This ranking is valid for all modules.

When selecting voting logic in redundant configurations it should be noted that there is a direct conflict between safety performance and production unavailability cost. For duplex modules, for instance, the 1oo2 voting logic gives the highest safety at the expense of an increase in production unavailability cost. The 2oo2 voting logic gives low production unavailability cost at the expense of a reduction in safety performance.

The improvement in reliability by employing physical replication of modules (hardware redundancy) is moderate, only. This is because the overall effect of hardware redundancy on system reliability is very much dependent on failure cause. Hardware redundancy is very effective against natural aging failures (independent failures). The technique is not that effective against failures due to excessive environmental stresses and human-made mistakes during design and operation; failures which often cause all branches to fail in the same manner (dependent failures). Thus, system design should be aimed at achieving the highest degree of independence between redundant modules. Also diverse redundancy might be a very effective way of improving the reliability.

It should be stressed that the results for *input/output* cards in Fig. 2 are valid for the specific case that there is a single sensor, single CPU and single actuator only. Other sensor/CPU/actuator configurations are considered in the following.

BALANCED CONFIGURATION - INPUT/OUTPUT CARDS

Figures 3 and 4 clearly show that the optimal I/O configuration is very much dependent on the voting logic of the corresponding field devices (sensors/actuators) and the CPU(s).

Figures 3a and 3b present results for input-configurations in systems with single and duplex CPUs, respectively. Figure 4 presents results for output-configurations with duplex CPUs. For each sensor (actuator) voting configuration (indicated on the top of the figure) safety calculations are presented for the relevant input(output) card voting configurations (illustrated in the margin to the right). The total configuration is obtained by inserting the input (output) card configuration into the sensor (actuator) configuration. The values of the critical safety unavailability refer to the contribution from the input (output) cards only. For instance, Fig. 3a shows that the CSU of single input cards in a configuration with two sensors (1oo2 voting) and single CPU is 10^{-5}. If sensor signals are distributed to different input cards (1oo2 voting), the overall CSU of the input cards is reduced to 10^{-6}. Thus, having 1oo2 voting for sensors (distributing these on different input cards) there is no need to include input card redundancy to improve safety. On the contrary, 2oo2 voting might in this case be used for input cards to improve LCC, without seriously affecting the safety.

Likewise, the chosen voting of actuators have direct implications for the choice of output card configuration. For example, having 1oo3 voting for actuators, distributing these on different output cards is a very safe configuration, and there should be no need to include output card redundancy to improve safety. On the contrary, 2oo2 voting might in this case be used for output cards, without seriously affecting the safety.

The figures illustrate that proper distribution of sensor signals on different input cards ensures also redundancy on the input cards.

SYSTEM SIZE AND TASK

A cost effective configuration for a process safety system is very much dependent on installation size; especially the number of areas that need protection.

When making safety requirements, the *number of safety functions* to be carried out upon an abnormal operating condition becomes essential. In offshore productions one safety function may for instance be to shut in one well. (This function may be carried out automatically by one out of three safety valves: Down hole safety valve, Master valve or Flow wing valve). Thus, an installation with say 12 wells will have 12 safety functions.

The contribution from output card/actuator to safety performance is approximately proportional to the number of safety functions. Thus, if the number of wells to be shut in upon an abnormal operating condition is high, the contribution from output cards/actuators tends to dominate the contribution from sensors, input cards and CPU.

This is illustrated in Fig. 5 which shows the contribution from output cards to Critical Safety Unavailability compared to the contribution from a single CPU with and without self-test. Note that the reliability of actuators are not included in the figure. The three straight lines for output cards correspond to three different voting logics (1oo1, 1oo2 and 1oo3). It is seen that the use of two or three valves per well implies that the contribution to loss of safety from output cards is less than the contribution from the CPU, even when the number of wells is as high as 100.

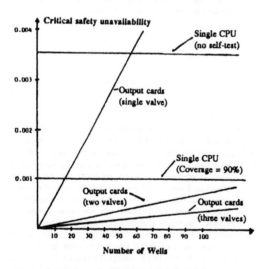

Fig. 5. Contribution to CSU from CPU and output cards for different number of wells to be shut in and valve redundancy.

SAFETY REQUIREMENTS FOR SUB-SYSTEMS

According to the results in Fig.s 3 and 4, the choice of input/output card redundancy must be related to the chosen sensor/actuator redundancy and CPU redundancy.

One practical way of handling this is to start by specifying the configuration of field devices and CPU. This can be done using Fig. 2. (Note that the results for input/output cards in Fig. 2 are only valid for the specific case that there is a single sensor/actuator and a single CPU only.)

Then the various input card configurations must be considered, given a particular sensor/CPU configuration. This can be done using figures like Fig. 3.

Next, the output card configurations must be considered, given a particular actuator/CPU configuration. This can be done using figures like Fig. 4.

CONCLUSION

Hardware redundancy on input/output cards gives only a moderate *overall* improvement in safety performance. The corresponding sensors/actuators contribute significantly more to safety unavailability than input/output cards do. Thus, *single* input/output cards and *redundant* sensors/actuators are generally an optimal configuration from a safety point of view. Proper distribution of redundant signals to different cards ensures also redundancy on the input/output cards. Double input/output cards with 2oo2 voting is only cost effective when cost per trip is high.

The fact that the CPU is demanded by all types of abnormal operating events (gas leak, fire and overpressure) and that a failure of the CPU may prevent any actuator to carry out intended safety action, often implies that the CPU should be duplicated. Two CPUs with 1oo2 voting and self-test is optimal when there is low/medium cost per trip. Two CPUs with one CPU in standby is also a feasible option. Three CPUs with 2oo3 voting and self-test should be considered when there are very high costs per trip.

REFERENCES

Aarø, R., L. Bodsberg, and P. Hokstad (1989). Reliability Prediction Handbook: Computer-Based Process Safety Systems, SINTEF report STF75 A89023. Trondheim.

Lydersen, S. and R. Aarø (1989). Life Cycle Cost Prediction Handbook: Computer-Based Process Safety Systems, SINTEF report STF75 A89024. Trondheim.

Hokstad, P., and L. Bodsberg (1989). Reliability Model for Computerized Safety Systems. In Proc. 1989 Annual Reliability and Maintainability Symposium. The Institute of Electrical and Electronics Engineering, New York. pp. 435-440.

Bodsberg, L., O. Ingstad, and T. Sten (1987). Alarm and shutdown frequencies in offshore production, SINTEF report STF75 A87010. Trondheim.

Poucet,A., A. Amendola,and P.C. Cacciabue (1987). Common Cause Failure Reliability Benchmark Exercise., EUR 11054 EN Commission of the European Communities Joint Research Centre, Ispra.

ACKNOWLEDGEMENT

This paper is based on work carried out by SINTEF within the research project "Reliability and Availability of Computer-Based Process Safety Systems". The project has been sponsored by the Royal Norwegian Council for Scientific and Industrial Research (NTNF), BP Petroleum Development Norway, EB Industry & Offshore, GP-Elliott Electronic Systems, Honeywell, Norfass, Norsk Hydro, Norske Shell, Phillips Petroleum Company Norway, Saga Petroleum, Siemens, Simrad Albatross and Statoil.

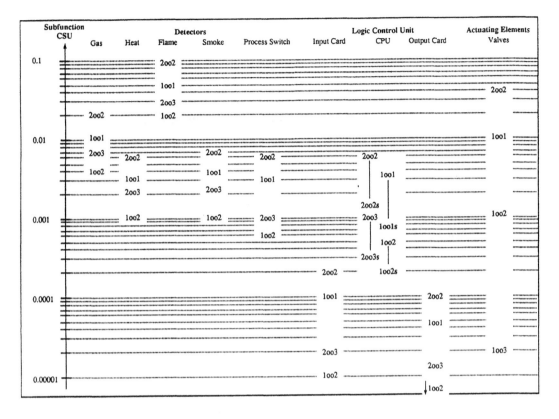

Fig. 2. Critical safety unavailability of safety system modules.

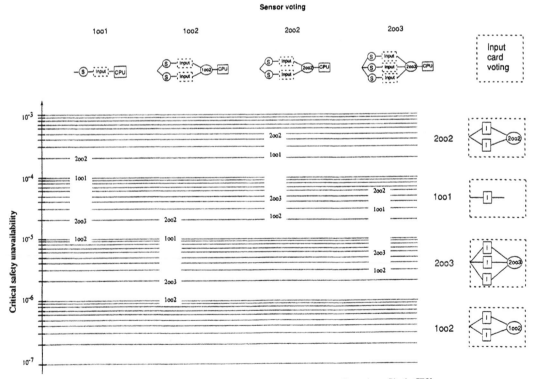

Fig. 3a. CSU of input card configurations for various sensor configurations. Single CPU.

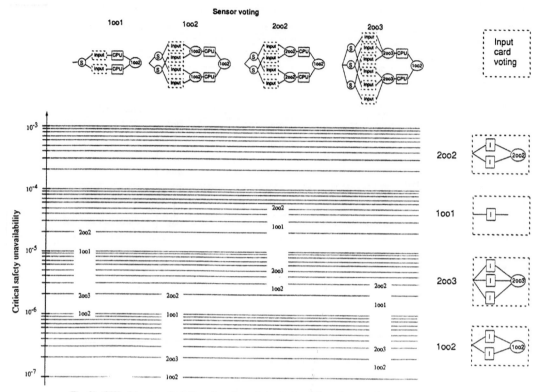

Fig. 3b. CSU of input card configurations for various sensor configurations. Duplex CPU, 1oo2 voting.

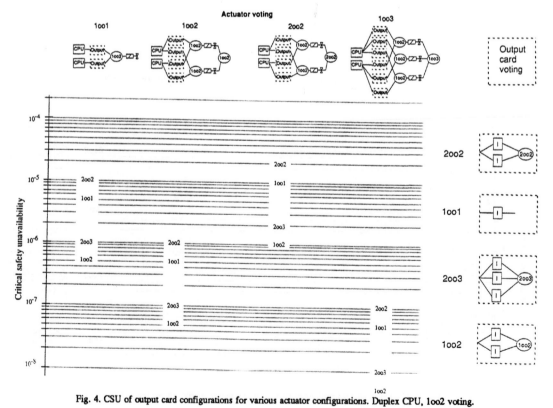

Fig. 4. CSU of output card configurations for various actuator configurations. Duplex CPU, 1oo2 voting.

COMPLEXITY MEASURES FOR THE ANALYSIS OF SPECIFICATIONS OF (RELIABILITY RELATED) COMPUTER SYSTEMS

C. Ebert* and H. Oswald**

*Institute for Control Engineering and Industrial Automation, University of Stuttgart,
Pfaffenwaldring 47, W-7000 Stuttgart 80, Germany
**Corporate Research and Development, Landis & Gyr Zug AG, CH-6301 Zug, Switzerland

Abstract:

Measurement has become recognized as a useful way to plan, control and evaluate software projects. Especially the early stages of the software development process, such as requirement analysis and design, are vital for the successful implementation of computer systems. High order Petri nets are introduced as a powerful formalism for the specification and analysis of concurrent systems. The paper describes the application of several complexity measures for Petri nets and their use to find overly complex structures. The assessment and evaluation of computer systems for reliability is illustrated with an exemplary computer controlled pay phone that was specified and analyzed with the Specs Petri net tool. Problems resulting from software/hardware interaction and different, however interacting, states of the system and their formalization with Petri nets are discussed. This technique supports the early analysis of sufficiently refined specifications that can be used as a system prototype for customer presentations and reliability assessment.

Keywords: *Complexity measurement, executable specification, Petri nets, process control, Specs, requirements specification.*

1. Control of Safety-Related Projects

Computer systems in today's society require an increasing amount of embedded software and hardware systems. Their use has introduced new problems for the software engineer combined with the complexity of such systems. As the costs associated with the development of such computer-based systems continue to fall, industry worldwide is increasingly coming to depend on new technology to maintain a competitive edge, while being able to operate ever more complex control systems that require a high degree of reliability, dependability and safety. Although safety-related systems have been around in nuclear power plants or in aeronautics for more than three decades, the growing use of computers to control such processes just recently raised the question whether they can reliably fulfil all specifications they were build for. Today even the specifications and their complete assessment is considered doubtful. As systems become more complex, a more critical appraisal of the whole development process is necessary: from the specification of what the system is required to do, through to the system design phase and the implementation of hardware and software components. The ensurance of safe system operation becomes the crucial point over a project's entire life time.

Measurement has become recognized as a useful way to plan, control and evaluate software projects. Especially the early stages of the software development process, such as requirements analysis and design, are vital for the successful implementation of computer systems. Errors and poor decisions made during these early stages result in costly and often intractable problems.

Historically, research in software measures has been done mainly in the area of source code. Current research in the area of complexity measures is directed towards using measures earlier in the life cycle. While the effects of controlling software reliability with complexity measures are well-known for source code, the influence of measuring are yet uncovered in other fields, such as the early specification and design phases or prototypes of mixed hardware and software components which are characteristic for real world systems.

Since unusually high complexity measures of any distinct part of the system's design might be caused by overly complex requirements, its early detection could be used to improve requirements as well. Different types and structures of specification and design documentation request a formal method to bridge this gap. This method was found in high order Petri nets [Genr87, Pete81] that are used to specify the behavior of hardware and software systems. A tool called Specs [DGGK87, OEM90, BEM90] has been developed to construct, simulate and animate such requirement specifications. The properties of high order Petri nets (e.g. hierarchical ordering with subnets, predicate transition nets with coded inscriptions defining guards and actions, arcs for dataflow, and tokens with specific values) permit complete and structured graphical modeling of requirements. Such Petri nets can easily be translated to design languages, hence bridging the gap between requirements and design.

2. High Order Petri Nets and Extensions

Literature on system design techniques includes a vast amount of different methods to specify software systems. Most methods focus on distinct aspects of the development process (certain parts of the process, distinct objects and their relations, etc.) and therefore they are only useful for distinguished application areas. High order Petri nets can be applied to the specification of the structure and the behavior of any kind of systems, even to entire embedded systems that require a tool with dynamic and parallel features. Environments of such systems usually include various technical components and many different man-machine interactions. Application areas are for example industrial control systems or office automation systems. The complete description of both computer control and environment behavior requires a complete formal model of these interacting subsystems. To analyze and prove the correctness of software components it is extremely useful if their descriptions and specifications take advantage of the same formal technique [Reis88].

Petri nets provide such a formal description technique because their general concepts of "events" and "conditions" can be applied to software and hardware components on the one hand, and human interactions on the other hand. High order Petri nets are not as well known as other graphical specification techniques such as data flow diagrams, though data flow diagrams only show static aspects of a system in contrast to high order Petri nets that in addition describe dynamic behavior. The combination of formally defined Petri nets with informal specification methods in order to provide formal semantics was showed by Pulli et al [PDGK88]. Results of the Petri net theory are analysis and verification methods, e.g. algorithms for proving invariants, detecting deadlocks and constructing reachability graphs. Current research in Petri net tools is directed towards executable specifications.

The Specs tool supports a subset of predicate transition nets (PrT-nets) [Genr87]. A simple PrT-net is given in Fig. 1. PrT-nets provide the following graphical notations: circles for predicates, boxes for transitions, arcs for data flow between those objects, and dots for tokens. A textual inscription can be assigned to each graphical symbol. Inscriptions at the arcs restrict their individual data flow, define conditions between several arcs ending at the same transition, and define names usable in the transitions. Transition inscriptions define guards and actions, where guards restrict the firing of transactions, while actions are executed when transitions are fired. Predicate inscriptions define initial tokens. Predicates contain tokens and their values during the simulation. The set of all tokens is the system state that changes with each simulation step. All other net elements remain unchanged during simulation.

Simulation and animation are based on the following firing rules: When data tokens are available, the arc conditions and the guard of all input predicates are evaluated and the output predicates are checked for spaces for tokens. When these conditions are fulfilled the transition may fire. After firing, all input data tokens are consumed and output tokens are sent to appropriate output predicates. The left side of Fig. 1 shows a net with a firable transition, while the right side shows the same net after firing.

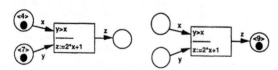

Fig. 1:   Predicate transition net with activated transitition (left side) and after firing (right side).

Large problems require methods for deviding them into smaller and less complex subproblems. It is hence impossible to describe large systems' specifications with Petri nets without a method for partitioning them. The theory of hierarchical refinement of Petri nets has already been described [Reis88]. The structuring elements are channels and agencies or instances. A channel is represented in Specs as a shaded circle, an agency or instance as a shaded box, both representing a number of elements of the next lower level of the hierarchy (Fig. 2, 3). A channel has the characteristics of a predicate on its level of appearence, while an agency resembles a transition. Channels and agencies might be nested, thus allowing several levels of hierarchical refinement. Such refinements can be defined top-down or bottom-up with the Specs tool.

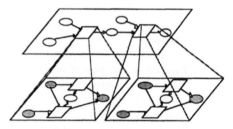

Fig. 2:   Net hierarchy showing a parent net with two subnets.

The Specs tools are built around a graphics based editor, simulator and simulation engine [DGGK87]. They permit the construction and animation of a hierarchically structured PrT-net by providing a net browser, an icon browser, an I/O browser and a function browser [OEM90]. The simulation engine allows the execution of a specification of an embedded system in real time in its real hardware environment. Four different execution modes are available: stepwise execution, animation with token movement, execution on a workstation, and fast execution on a simulation engine built around a series of transputers [BEM90]. Main problems concerned with fast simulation are transforming Petri nets into Occam processes, generating Occam code, distributing the processes optimally on the available processors, loading down code and data, and loading up data for presenting token distribution in the net browser.

3. Early Life-Cycle Measures for Petri Nets

In the field of software measures we distinguish between process measures and product measures [CDS86]. To avoid any misunderstandings, we define measures as the methods of associating any input to a number with respect to order, scale, and range. Examples for process measures are given in [GrCa87], classic examples for product measures are described in [HSYT90]. In this context, we are refering to product measures.

While applying the stepwise refinement process, it was revealed that complex iterations between requirements and actual design often hinder all managerial instruments for control. Design is connected to all steps originally considered independent. Requirements analysis results in the application design and eventually in the system design; the so-called preliminary design and the detailed design result in functional and logical design; coding, implementation, and testing result in linguistic and behavioral design; and maintenance results in renovative design [Curt90]. Management of the design process obviously requires a better understanding of requirements specification and its contribution to the phases of the process.

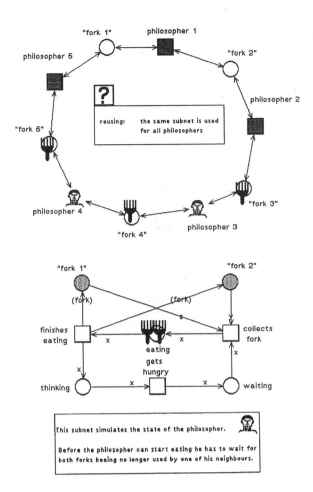

Fig. 3:   Top net and one subnet with examples of comments, icons and buttons.

Current research in the area of complexity measures is directed towards using them earlier in the life cycle. This resulted in measures based on program description in pseudo code. To apply complexity measures as early as possible in the life cycle we defined a series of measures that are based on the formal specification based on PrT-nets as described earlier. By using measures in different phases (e.g. specification measures; design measures for the preliminary design; design language based measures for the detailed design; and code complexity measures for the program code), it should be possible to trace complexity inherent to distinct requirements troughout the software development process and to separate it from additional solution-dependent complexity. Therefore, the individual accountability of distinct system design decisions and their impact on quality or productivity requirements can be traced.

A number of complexity measures are available for the Specs tool. They can be applied to a single subnet or to each subnet of a complete net hierarchy. These numbers are provided in a prompter or in a different text window (Fig. 4). The following complexity measures have been implemented: **cyclomatic number** (based on McCabe's cyclomatic complexity measure [CDS86] extended to concurrent programs [DeMo90] according to the formula:

$$v(Pn) = F - T - S + 2*K,$$

where F is the number of arcs, T is the number of T-Elements, S is the number of S-elements, and K is the number of separate components); **fan-in** (maximal number of arcs ending in one distinct element); **fan-out** (maximal number of arcs departing from a distinct element); **T-elements** (number of T-elements showed in the distinct subnet excluding agencies); **S-elements** (number of S-elements showed in the distinct subnet excluding channels); **arcs** (number of arcs in the subnet); **channels/instances** (number of channels and instances or agencies showed in the subnet); **interfaces** (number of interfaces or number of S- or T-elements defined in an upper level of the hierarchy); **separate components** (number of separate components or programs of the subnet).

By evaluating the different measures of an individual net simultaneously it is possible to rate its complexity. The number of subparts or elements of a given net is the first hint to its overall complexity. As subparts are not always of the same individual complexity they should be investigated individually. A high number of channels and instances represents components that are further refined, thus contributing to complexity on a lower level. Basically, the refinement of elements usually helps in decreasing the complexity (as perceived visually) of a distinct level of the hierarchy. T-elements should be considered more intensive than S-elements because the latter ones just represent data and translation of data, while T-elements might contain complicated inscriptions (conditions). Fan-in and fan-out is a pictorial data that describes the number of arcs starting or ending at a single component, therefore measuring interface complexity. Both should be kept as low as possible.

Another form of complexity, not yet considered, is what we perceive visually while dealing with PrT-nets. The number of intersections of arcs, their lenghts, the distances between connected elements, their relative position or alignment are factors that highly contribute to the complexity we intuitively attach to a net. Further research of visual complexity should help to improve the graphical representation.

97

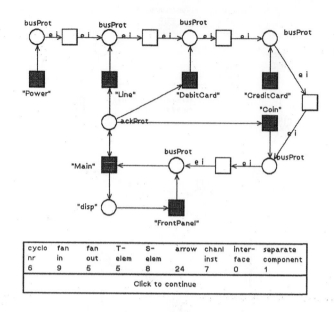

cyclo nr	fan in	fan out	T-elem	S-elem	arrow	chanl inst	inter-face	separate component
6	9	5	5	8	24	7	0	1

Click to continue

Fig. 4 a: Top net of *Payphone* with complexity measures (respecified version).

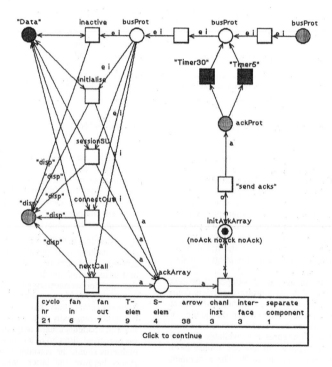

cyclo nr	fan in	fan out	T-elem	S-elem	arrow	chanl inst	inter-face	separate component
21	6	7	9	4	38	3	3	1

Click to continue

Fig. 4 b: Subnet *main* of Payphone with complexity measures (respecified version).

4. An Application: The Payphone Project

As an example for the application of complexity measures to the specification with Petri nets a public payphone station is described and analyzed in this section. The payphone station has been specified for a customer with Specs by the Landis & Gyr Corporation. Some features making the project difficult are several payment modes, a small display for messages to the user and a keypad for number dialing and special services. Payment modes supported by the station are cash (coins), credit cards and optical debit cards. The display shows messages about the current state of the payphone, e.g. all digits of a partially entered phone number. Any manipulation error is also commented on this display. Additional buttons provide user shortcuts, e.g. for entering a complete emergency number by pressing a single button.

98

The specification for this project was done with the Specs Petri net tool for several reasons:

- specification of the man-machine interface;
- application of this specification as an executable prototype;
- support for the validation of the specification by the user;
- automatic generation of documentation;
- generation and validation of Input/Output test sequences.

A specification for this public payphone as well as an executable prototype were developed with the Specs tool. Since the customer had provided an executable Hypercard/Hypertalk textual specification, it was fairly easy to start with the more formal oriented Petri nets approach. The Petri net specification graphically describes the logics of the payphone and the complex dynamic relations between its components. The executable, however still workstation-based, prototype of the payphone provides an early view as it will be presented to future users, thus allowing to test and validate the complete functionality as early as during the specification process. All displays and buttons are presented on the screen. User interactions can be triggered with the mouse and the results are presented immediately on the screen. Various experiences have been made while specifying with Petri nets. They are described in the rest of the article.

With Specs it was possible to locate some errors. Many state sequences have been found which are missing in the customer's specification as well as in his Hypercard prototype. Most of the missing state sequences are located in the rather complex man-machine interactions. Modeling with Petri nets has advantages while specifying time behavior. Most of the faults and inconsistencies to the original textual specification thus were linked to messages that need to be showed on the display for a predefined time. Finally, the complete specification with Specs has been informally accepted by the customer as a correct specification of the man-machine interface. The formal contract is based on a test-sequence document. This document has been checked against the specification by executing and animating the test-sequences with the Specs tool.

The results of analyzing the net hierarchy and its complexity is given in Fig. 5a. Each subnet was measured separately as indicated by the several lines of the table. The complete specification showed to be relatively complex. Therefore two alternative solutions have been applied in order to reduce complexity. The approach of the second solution has been to reduce the number of elements. Fig. 5b shows its complexity measures. The structural complexity measures decreased considerably. However, complexity is moved to the textual inscriptions of the T-elements which is not yet considered in these measures. Since both alternative solutions proved not to permit any complexity reductions, it seemed as if the overall complexity originates from the functionality requested by the customer and less from individual approaches for the solution. This complexity hence must either be accepted with the implicit danger of software faults and potentially safety-critical states, or functionality must be reduced or at least simplified. Another approach for handling designs with high overall complexity would be to supply more test time and enhanced test plans considering much more states of the payphone as forseen.

Time spend for the specification (with Petri nets as described):
  first specification: 2.5 person months (within 7 months), complete
  2 respecifications: 1.5 person months (within 2 months), incomplete (70%, 20%)

Size of first specification:

T-elements	408
S-elements	248
Arrows	2507
Subnets	100

global complexity:
  Fan In/Out: number of arcs in / out of an S- or T-element
  maximal Fan In of a single T-Element: 68 (very complex)
  maximal Fan Out of a single T-Element: 68 (very complex)

distinct measures for subnets

cyclo nr	fan in	fan out	T-elem	S-elem	arrow	chanl inst	inter-face	separ. comp.	net name
10	12	12	0	4	14	2	0	1	PayphoneTop
19	66	8	0	14	41	6	4	1	-Payphone
17	5	5	12	7	38	1	3	1	--DCRModuleIO
4	2	3	2	0	8	0	4	1	--eject
35	**68**	**68**	**0**	**11**	**66**	**4**	**18**	**1**	**--MainModule**
14	7	7	0	6	31	6	7	1	--KeyHandler
1	1	1	4	0	8	0	5	1	--dispatchKey
13	3	3	4	2	26	2	7	1	--dialingKeyMgmt

etc.

Fig. 5 a: Complexity measures for the payphone project (first spec.).

Size of the respecification:

T-elements	124
S-elements	84
Arrows	436
Subnets	26

global complexity:
Fan In/Out: number of arcs in / out of an S- or T-element
maximal Fan In of a single T-Element: 10
maximal Fan Out of a single T-Element: 10

distinct measures for subnets

cyclo nr	fan in	fan out	T-elem	S-elem	arrow	chanl inst	inter-face	separ. comp.	net name
6	9	5	5	8	24	7	0	1	PayPhone
1	1	1	4	2	6	0	1	1	-Power
4	10	10	3	2	10	1	2	1	-Line
14	4	4	10	6	30	0	2	1	-DebitCard
19	5	5	12	7	37	0	1	1	-CreditCard
9	3	3	8	5	22	0	2	1	-Coin
21	6	7	9	4	38	3	3	1	-Main

etc.

Fig. 5 b: Complexity measures for the respecified payphone project.

5.   Conclusions

High order Petri nets are used for developing and executing specifications of embedded systems. This technique permits a systematic approach for capturing requirements by executing the specification, a formal notation for expressing them, and a tool to assist in their validation. The advantages of this formal basis are the elimination of ambiguities during requirements analysis and the support to analyze, visualize and verify the specification early in the life cycle. Especially the application of Petri nets as an efficient formalism for specifying concurrent systems as well as user interfaces seems to be valuable for early and intensive user demonstrations and customer negotiations. The Specs tool provides the kind of environment needed to accomodate frequent changes and experimentations necessary during the specification and prototyping process.

This paper presents an approach to integrate software specification measures based on predicate transition Petri nets into the Specs Petri net tools. Since today's software engineering is dealing with the system's software and hardware development, it was necessary to select an environment supporting both interacting components. By quantifying aspects that make such specifications complex we are able to give system engineers helpful hints to improve their system design early in the life cycle. Finally an industrial application of the tool and the problems related to decreasing these complexity measures showed that there is a problem-inherent complexity that is almost constant and a solution-inherent complexity that might be reduced by applying different design methods or by refining requirements.

Bibliography

[BEM90]   Bütler, B., Esser, R. und Mattmann, R. : A Distributed Simulator for High Order Petri Nets. In: *Advances in Petri Nets 1990*, Hrsg.: G. Rozenberg. Springer Verlag, Heidelberg, 1990.

[CDS86]   Conte, S. D., Dunsmore, H. E. und Shen, V.Y.: Software Engineering - Metrics and Models. The Benjamin / Cummings Pub., Menlo Park, CA, USA, 1986.

[Curt90]   Curtis, B.: Modeling, Measuring, and Managing the Software Development Process. *Tutorial on the 12th International Conference on Software Engineering 1990*, IEEE Comp. Soc. Press, New York, NY, USA, 1990.

[DeMo90]   DePaoli, F. und Morasca, S.: Extending Software Complexity Metrics to Concurrent Programs. In: *Proc. of the IEEE COMPSAC 1990*. IEEE Comp. Soc. Press, Washington, DC, USA, 1990.

[DGGK87]   Dähler, J., P. Gerber, H.-P. Gisiger und Kündig, A.: A Graphical Tool for the Design and Prototyping of Distributed Systems. In: *ACM SIGSOFT Software Engineering Notes*, vol. 12, no. 3, pp 25 - 36, Jul. 1987.

[Genr87]   Genrich, H.: Predicate/Transition Nets. In: *Petri Nets: Central Models and Their Properties*. Editors: W. Brauer, W. Reisig and R. Rozenberg, pp 18 - 68. Springer Verlag, Heidelberg, 1987.

[GrCa87]   Grady, R. B. und Caswell, D. L.: Software Metrics - Establishing a Companywide Program. Prentice Hall, Englewood Cliffs, NJ, USA, 1987.

[HSYT90]   Hirayama, M., Sako, H., Yamada, A. und Tsuda, J.: Practice of Quality Modeling and Measurement on Software Life-cycle. *Proceedings of the 12th International Conference on Software Engineering 1990*, IEEE Comp. Soc. Press, New York, NY, USA, 1990.

[OEM90]   Oswald, H., Esser, R. und Mattmann, R.: An Environment for Specifying and Executing Hierarchical Petri Nets. In: *Proc. of the 12th Int. Conf. on Software Engineering*, IEEE Comp. Soc. Press, Washington, DC, USA, 1990.

[Pete81]   Peterson, J. L.: Petri Net Theory and the Modelling of Systems. Prentice-Hall, Englewood Cliffs, NJ, USA, 1981.

[PDGK88]   Pulli, P., Dähler, J., Gisiger, H.-P. und Kündig, A.: Execution of Ward's Transformation Schema on the Graphical Specification and Prototyping Tool SPECS. In: *Proc. of the CompEuro 1988*, pp. 16 - 25, Brussels, 1988.

[Reis88]   Reisig, W.: Embedded System Description Using Petri Nets. In: *Embedded Sytems*. Editors: Kündig, A., Bührer, R.E. und Dähler, J. pp. 18 - 62. Springer Verlag, Heidelberg, 1988.

CLASSIFICATION OF CRITICAL EVENTS IN SYSTEMS DESCRIBED BY GRAFCET USING THE MARKOV PROCESS

Z. Abazi and T. Peter

Laboratoire d'Automatique de Grenoble (E.N.S.I.E.G.),
B.P. 46-38402 Saint Martin d'hères, France

Abstract This paper presents a method based on Markov chains to model sequential processes described by grafcet. Thanks to a mathematical approach, this method allows the dynamic aspect of the system to be taken into account and a model of failure to be introduced for a given fault . One presents reduction rules when studying the grafcet with numerous states, and laws to construct the markovian model from the grafcet. By solving the corresponding Markovian model, we shall be able to classify the different failures which can affect the system by their appearance probabilities.

Keywords: Safety, Grafcet, Classification, Markov process.

I- INTRODUCTION

Safety is an important aspect of process control systems and must be considered in fault tolerant level. The grafcet is a well known modelling tool for sequential processes. Concerning more especially the industrial process, most of them (in particular the mecanical ones) are based on the grafcet tool. It is used either to design or model the control systems. Then one can say, all sequential processes can be modeled by grafcet. However, with this modelling tool, when breakdown occurs, this model becomes unserviceable. Therefore, the aim of this research is to take into account possible failures in the modelling level which enable us to provide for emergencies. Numerous works on this subject deal with different approaches based on the Petri nets controllers (Leveson,1987) and (Zhou, 1989). As the grafcet is a well known tool and mainly used to describe industrial processes, our study is based on this tool. After a synthesis of the several works in this field, a global approach enabling the study to be succesfully competed, is proposed. It is made up of four parts :

Part 1 : *Identification of faults*

The analytic methods are used to show off the set of faults. This part allows a panel of faults to be obtained .

Part 2 : *Modelling and Calculation*

For any fault of the panel, a method is defined which enables us to model and calculate some probabilities, especially the appearance probability of each fault.

Part 3 : *Classification and correction*

Knowing for each fault, its "critical degree" (part 1) and its appearance probability (part 2), faults are classified and a method defined to correct the grafcet.

Part 4 : *Validation*

For our purpose, it is assumed that for a system described by grafcet, one studies only a fault from the panel of faults obtained in part 1. This work then concerns part 2. The aim is to define a new method which lets us calculate and classify by the appearance probability for a given fault of the panel of faults defined in part 1. For this approach the Markov process is used as the mathematical calculation tool .

This paper is organized as follows : After a few words on the modelling tool used : the grafcet, section III presents the principle of the proposed method which is illustrated by an example. This method is made up of several reasoning steps which are defined in this section. The calculation of appearance probabilities for the faulty systems is detailed in IV. And finally, some concluding remarks end this paper.

II- THE GRAFCET

The grafcet is a modelling tool which can be considered as an application to study the coordination of asynchronous events. The formal definitions of the grafcet can be found in (Inter. Standart 1988) and the notation and teminology used in this paper closely follow that of (David,1990). Some definitions are given below :

A grafcet is made up of two kinds of nodes *places* and *transitions*. The directed arcs interconnect these nodes. A static representation is obtained with this set of places and transitions combined in this way. Due to the evolution rules, this representation also has a dynamic aspect corresponding to the evolution of the control process. Moreover, one or more *actions* may be associated with each place and one *receptivity* shall, be associated with each transition. At a given time, a place may either be active or inactive. When the place is active the corresponding actions are performed. The set of active places defines a *situation* of the considered system. One situation corresponds to the state of the system. A transition is firable if and only if all its input places are active and its receptivity is true.

Fig. 1-a presents an example of a grafcet with 4 places and 4 transitions. It corresponds to the behavior of a walker when he crosses over the road at the traffic light. Place 1 is active. The situation is represented by the set of active places (1). It corresponds to the initial situation. When the logical condition r1 is true, then transition t_1 is firable as place 1 is active. After firing this transition (t_1) another situation of this system is obtained : (2). The evolution continues as long as there exists at least one enabled transition.

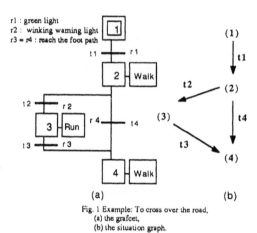

Fig. 1 Example: To cross over the road,
(a) the grafcet,
(b) the situation graph.

For a given initial situation, the **situation graph** of a grafcet is defined as the set of the situations which result from the different evolutions of the grafcet. To represent the situation graph, a graph is used where the nodes are the situations of the system labeled with the present active places and the arcs represent transitions between situations as shown in fig 1-b.

III- FROM THE GRAFCET TO THE MARKOVIAN MODEL

Each application can be described by Grafcet. To study part 2 of our project (modelling and calculation), starting from a grafcet, the following three steps are needed to model and evaluate the occurrence probability of each fault :

Step 1 : The situation graph is deduced from the grafcet. This step includes any reduction rules applied to the grafcet.

Step 2 : In this step, a markovian model is constructed from the situation graph. A method is also defined to reduce the markovian model when this one contains a large number of states.

Step 3 : For a given fault, the markovian model of the faulty system is defined in this step. The probability calculation is obtained by solving this markovian system.

For a better understanding, throughout the paper we use the same more realistic example shown below to illustrate the several steps of the method.

The truck example

Two trucks T_1 and T_2 are considered travelling between two extreme positions left (B) and right (A_1) for T_1 and (A_2) for T_2 see fig. 2-a. When truck T_1 is in A_1 and if the operator actuates push-button M1, the cycle begins:
- loading up of T_1,
- moving towards B,
- unloading of T_1,
- coming back to A_1.

Truck T_2 has the same cycle.

These two trucks have a common "frustum" delimited by Sr_i and B. If the frustum is engaged by one truck when the other is in Sr_i, this one waits for it to be released.

In the case of simultaneous request in Sr_1 and Sr_2 truck T_1 has priority. The grafcet of this system is given in fig 2-b .

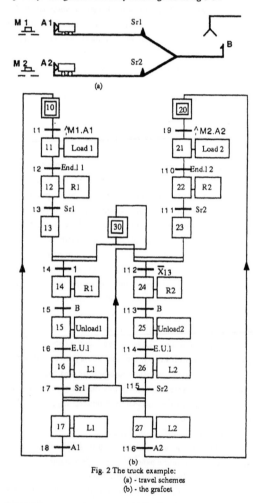

Fig. 2 The truck example:
(a) - travel schemes
(b) - the grafcet

Remark

The first part of our study gives a panel of faults where the critical events are the ones where trucks T_1 and T_2 are in the commun frustum. It is the case where from sensors Sr_1 and Sr_2 have broken down. Then in our study these sensors must be shown from the modelling point of view. From the calculation point of view, only these cases of faults are taken into account.

III-1 CONSTRUCTION OF THE GRAPH-MODEL

The grafcet modeling tool offers the possibility to start with an overall description of the control system and may, via progressive levels of description, proceed to a description in which all the details of the system are revealed. Our work is therefore based on a grafcet where all details of the system are revealed.

The outline of the proposed method is as follows : From the system modeled by grafcet, one deduces the situation graph composed by the reachable states of the system. In the truck example given in fig. 2, the situation graph is more complicated as it contains a large number of situations (52 situations and 98 arcs !). Meanwhile, when the grafcet concerned has a large number of places, one proceeds at once to reduce this one, then the situation graph is deduced. The reduction technique is defined in III-1. Starting from the situation graph, one state of Markov chain is associated to each situation. A Markov chain is then obtained called a **graph-model**. The construction of the graph-model is detailed in III-2. The main limitation of such modelling tool is the graph model complexity . This is why, we propose in section III-2.1 the simplification rules to the Markov chain deduced from the possibly reduced situation graph. This leads us to the reduction model called a **re-model** defined in section III-3.

III-1.1 Situation graph

The situation graph is deduced directly from the grafcet. It corresponds to reachable situations of the system. This construction method is classical and consists in studying all firable transitions from the initial situation. This process is repeated until either we reach a blocking situation (no firable transition) or we obtain one of the previous situations. We assume for this construction that all the receptivities are always true.

In practice, when the grafcet contains great state numbers and/or presents a parallel sequence, the situation number increases and leads to a graph model with a large number of states. For example fig. 3 represents a grafcet with m linear sequences and each sequence contains n places. This representation involves a graph model with n^m situations which gives a complicated model. Therefore, before constructing the graph model, we propose the reduction rules applied to the grafcet.

III-1.2 The simplification rules of the grafcet

The only way to reduce the state number is to associate a single state with set (place, transition) of the grafcet. The reduction rules allow the reduced grafcet to be obtained. They refer to the "linear sequence " defined as follows.

Definition 1

A **linear sequence** Si is made up of a series of places {m,n,...} which will be activated one after the other. In this structure, each place is followed by only one transition and each transition is firable by one place.

e.g. There are four linear sequences in fig.2-b:

$$S1=\{17,10,11,12,13\},$$
$$S2=\{27,20,21,22,23\},$$
$$S3=\{14,15,16\},$$
$$S4=\{24,25,26\}.$$

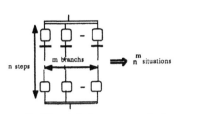

Fig. 3 Example leading to numerous situations.

Reduction rule

To obtain the reduced grafcet from the grafcet of the whole system, a place Si is associated with each linear sequence .

But two reasons may stop the reduction :

1) the complexity of the reduced grafcet is considered as low enough (voluntary stopping),

2) the reduced grafcet is irreducible if no linear sequence exists.

The user must thus reach a compromise between complexity of the reduced grafcet choosen to model the system and information we want to show off when the failure occurs.

III-1.3 Application to the truck example

In our example we have in fig.4-a the corresponding reduced grafcet (it is a maximum reduction according to our rule). There are a minimum of places but also a minimum of information. It is not possible to distinguish between the position of truck T_1 (resp. T_2) on A_1 (resp. A_2) or the position on Sr_i in the places S1 (resp.S2).

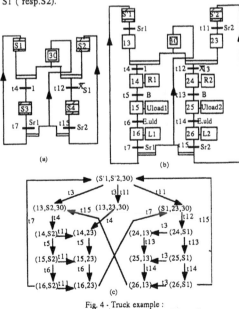

Fig. 4 - Truck example :
(a) Maximum reduction of the grafcet,
(b) Reduced grafcet,
(c) Situation graph.

In this example as explained above, the interesting case to study is the one where failure leads to the two trucks in the common frustum . This can be caused by the failure of sensors B or Sr_j. Then in the reduction of the grafcet we must show off sensors B , Sr_1 and Sr_2. That is why, from the grafcet in fig.2-b, one state S'_1 is associated to a linear sequence {17,10,11,12} and S'_2 to {27,20,21,22}. The reduced grafcet in fig. 4-b is then obtained.

The situation graph associated to the reduced grafcet of fig.4-b contains 16 situations (instead of 52 in the first one) see fig.4-c. This first reduction enables us not only to reduce the state number of the situation graph, but also to focus the study on the particular faulty state.

III-2 MARKOVIAN MODEL

The Markov process considered in the method is the discrete Markov system with discrete time. The transition probabilities are then used in the calculation.

III-2.1 Principle

The graph model is obtained directly from the situation graph. To describe all situations of the considered system, a flow graph representation is used. The nodes of the directed and labeled flow graph represent each situation of the situation graph deduced by the grafcet. There are as many nodes as there are situations. The node which represents the situation (i,j,k,...) is denoted by M followed by an evocative abbreviation of the state of the system.

e.g. : in fig. 5-a,

$M1s$ means that truck T_1 is in the common frustum knowing that T_2 is on standby ;

$M1l$ means that truck T_1 is in the common frustum knowing that T_2 is loading up.

□

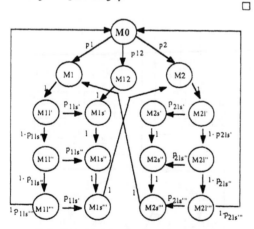

Fig.5-a Markovian model of the truck example: Graph-model.

There is a directed arc from node Mix to node Miy if and only if the system is in the situation y after x and the corresponding label denoted p_{ixy} gives the probability to be in Miy after Mix ($0 < p_{ixy} \leq 1$).

e.g.: in fig. 5-a,

p_{11s} is the probability that truck T_2 finishes loading up knowing that T_1 is in the frustum. It is the probability to move from M_{1l} to M_{1s}.

□

By this flow graph, a markovian model is obtained called a **graph-model** (Markovian model from the situation graph) which gives the model of the system behavior. If the situation graph contains n situations, the graph-model is then the Markov chain with n states.

The graph-model of the truck example defined above, contains 16 nodes as shown in fig. 5-a.

III-2.2 Reduction of the markovian model

The graph model can be reduced by gathering some states of Markov chain in only one "macro-state". This reduction facilitates the calculation. It depends on the fault to be studied and is applied directly on the considered application.

Application to the truck example

The main interest are sensors Sr1 and Sr2. It matters little to know whether truck T_1 (or T_2) is in loading up position or on standby in Sr_1 (or Sr_2) when the other one is in the common frustum. Then, a single macro-state represented by a double circle, is associated to the set of states such that :

M_{1l} is associated to { $M_{1l'}$, $M_{1l''}$, $M_{1l'''}$ },

M_{1s} " " " { $M_{1s'}$, $M_{1s''}$, $M_{1s'''}$},

M_{2l} " " " { $M_{2l'}$, $M_{2l''}$, $M_{2l'''}$ } and

M_{2s} " " " {$M_{2s'}$, $M_{2s''}$, $M_{2s'''}$}.

A reduced graph-model is obtained denoted **re-model**, see fig. 5-b with only eight states.

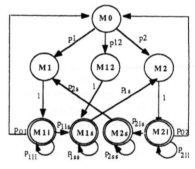

Fig. 5-b Markovian model after reduction : re-model.

III-2.3 Transition probability

Let Pr[i/GM] and Pr[i/RM] denote the stationary probabilities of state i in the graph model and re-model respectively. More formally, the notion of equivalence is defined as follows.

Definition 2

The graph-model and re-model associated with a system are equivalent if and only if the state probabilities are such that:

1) for each state i represented by a circle in the graph-model,

$$Pr[i / GM] = Pr[i / RM] \text{ and,}$$

2) for each macro-state M_i,

$$Pr[M_i / RM] = S \, Pr[j/GM]. \qquad \square$$

By construction, most of the transition probabilities in the re-model are deduced directly from the graph-model. The probability to move from state M_i in the re-model is $1-t_i$, where t_i is the probability that the system remains in M_i. t_i known as the reduction coefficient associated with M_i is given by the following theorem which has been formally proven in (Peter,1990). This theorem refers to fig 6 which represents the general case of the reduction.

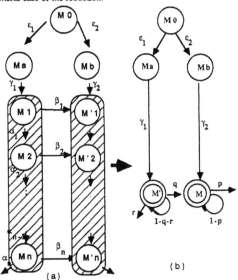

Fig. 6 The successive reductions
(a) global markovian model
(b) markovian model after the reduction :
reduction coefficients : p, q and r

Theorem

The graph-model and re-model (fig.6) are equivalent if and only if :

$$p = \frac{\gamma_2.\varepsilon_2 + \gamma_1.\varepsilon_1 \left(1 - \prod_{i=1}^{n} \alpha_i \right)}{n \, \gamma_2.\varepsilon_2 + \gamma_1.\varepsilon_1 \left(n \sum_{i=1}^{n} \left(\prod_{j=1}^{n} \alpha_j \right) \right)}$$

$$q = \frac{1 - \prod_{i=1}^{n} \alpha_j}{1 + \sum_{i=1}^{n-1} \left(\prod_{j=1}^{i} \alpha_j \right)} \quad ; \quad r = \frac{\prod_{i=1}^{n} \alpha_j}{1 + \sum_{i=1}^{n-1} \left(\prod_{j=1}^{i} \alpha_j \right)}$$

with n = state number in M and M',

$a_i = Pr[M_i \rightarrow M_{i+1}]$; $b_i = 1 - a_i$;
$g_1 = Pr[M_a \rightarrow M_1]$; $g_2 = Pr[M_a \rightarrow M'_1]$;
$e_1 = Pr[M_0 \rightarrow M_a]$; $e_2 = Pr[M_0 \rightarrow M_b]$. $\qquad \square$

Particular case

For $n = 1$, the theorem gives $p = 1$, $r = a_1 /(1 + a_1)$ and $q = (1 - a_1)/(1 + a_1)$, which is an expected value.

III-2.4 Numerical results

The reduction formulas are applied to different tests. Among them, we propose the case where, truck T_1 is more used than T_2.

For each test, fig. 7 presents the stationary states of Markov chain for two models :

a- without reduction (for the graph-model deduced directly from the grafcet)

b- after the grafcet reduction (markovian model deduced from the re-model.

Prob states	Graph model	re model		Prob states	Graph model	re model
M11'	0,1476			M1s'	0,0084	
M11"	0,1402	0,4140		M1s"	0,0225	0,0723
M11'"	0,1262			M1s'"	0,0414	
M21'	0,0438			M2s'	0,516	
M21"	0,0508	0,1461		M2s"	0,0077	0,0601
M21'"	0,0515			M2s'"	0,0008	

Fig. 7 Numerical results

These numerical results show the equivalence between the different models.

IV- MARKOVIAN MODEL FOR THE FAILED SYSTEM

In the case of the system described by grafcet, a faulty sensor can cause a catastrophic state of the system. We are particularly interested in faults involving a critical event. This section then deals with time evolution of the failures. To describe the failed system, the common representation is used as shown in fig.8. On the one hand the failed system is represented by the set of states RF (Right Functioning). It represents the system in good running order. On the other hand, the second one WF (Wrong Functioning) represents this system in irregular operation. When the failure occurs, the system can reach an absorbing state D (Detection) from the set of states WF. The appearance probability is obtained by solving this Markov system and corresponds to the probability of reaching the absorbing state.

IV- 1 Fault models

A control system is made up of two parts : an operative part (O.P.) and control part (C.P.). The operative part is made up of the process. The control part thus gives orders to the operative part. A control system is made up of a set of fault F = C ∪ O with C = { fault in the Control part} and O = { fault in the Operative part} . To study the dependability of the process, the operative part is studied . It is for example the case where a machine fails or a positioning sensor breaks down.

IV- 2 Appearance probability calculation

An error may cycle between its latent and effective state and , in general, propagates from one situation to another. By propagating, an error creates other new errors which in turn may propagate or remove, and so on. To take into account the two possibilities, propagation or removal, the state RF has to be doubled to WF. Hence, if m is the state number of RF, a 2m+1 state Markov chain is needed to describe faulty system behavior. Fig. 8 shows the general model suitable for any fault, and possible transition between states. The system reaches the state WF with the probability P_a when the system has failed. P_b is the probability that the system will start to run properly again (the fault is masked). The system moves to state D with probability P_d when the failure is detected.

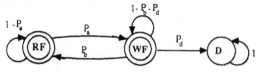

Fig.8 Markov chain describing the general behavior of faulty system.

Therefore, each fault f requires an appropriate model denoted FM_f (Failed Model for the fault f) in which states RF and WF are represented by several states deduced from the re-model of the system. Let $P_D(t)$ be the probability that FM_f is in state D after t steps. The steps number T_f needed in order to detect f with a given probability Q_D is equal to the lowest value of T_f such that $P_D(t) \geq 1 - Q_D$.

For a given fault f_i, three set of states are considered called "set of faults" : A, M and Dt where,

A is a set of states which can be affected by this fault,

M is a set of states which can mask the error,

Dt is a set of states which can detect the error after t steps.

IV- 3 Application to the truck example

Each fault requires an appropriate model. Let us consider a fault f_1 which affectes a sensor Sr1. The construction and calculation of this model for the fault f_1 will be detailed below. The set of faults are the following:

A = {Mo, M1, M12, M11, M1s},
M = {M1, M12, M11, M1s},
Dt = {M2, M21, M2s}.

Knowing these three sets the failed model FM_{f1} can be constructed for the fault f_1. Then each of the states RF and WF is replaced by eight states of the re-model. Transition RF--->WF is brought out by the transition from the set of states A to the following one with a failure probability P_a and the transition WF ----> RF by a set M with masking probability P_b. State D is reached from the set of states Dt with the detection probability P_d. The model FM_{f1} is then obtained. The system solution leads to T_{f1} = 5649 steps for QD = 0.95. 5649 steps can be said to be needed to reach a probability equal to 0.95 for the absorbing state D.

Numerical results

The following table gives the results obtained by solving the markov model in the case of fault f_1 (resp. f_2, f_3) when sensor Sr1 (resp. Sr2, B) has broken down. This indicates the step number T_f enabling us to reach the detection state D with a probability equal to 0.95. These results concerne the case where truck T_1 and T_2 are equaly used.

f	T_f
f_1	5649 *steps*
f_2	5665 *steps*
f_3	5655 *steps*

Numerical results obtained for different faults and different probabilities are detailed in (Peter,1990). The paper gives only one of them .

These results show that :

1) For every test the result is constant for sensor B,

2) For Sr1 and Sr2 the result depends on the using probability,

3) The priority for T_1 has a little weight but discernible between T_1 and T_2.

This model enables us to classify the fault according to the occurrence probability by comparing the probability of the state D in the model for the same detection quality. Then it can be said for this example that a failure in sensors Sr1 can be revealed before Sr2. It is obvious that sensor Sr1 is more used than Sr2 in so far as Sr1 has priority than Sr2 . This explains the little difference between these two values T_{f1} = 5649 (for Sr1) and T_{f2} = 5665 (for Sr2).

V- CONCLUDING REMARKS

Thanks to the proposed modelling tool, the occurrence probability for different faults can be calculated by solving a Markovian model associated with the faulty behavior. We are currently developing a complete routine to computerize the method and which takes into account the modelling tasks. Starting with the flowgraph description, the routine gives the transition matrix, then for any fault f, it constructs the FM_f model. The program computes its transition probability of a particular state or the steps number enables us to reach the state D with a given probability.

References

David, R. and Alla , H.(1989). Du grafcet aux réseaux de Petri . Edition Hermes

International Standard (1988). Etablissement des diagrammes fonctionnels pour les systèmes de commande. IEC- 848- first edition .

Leveson, N.G. and Stolzy J.L.(1987) Safety analysis using Petri nets IEEE Trans. Soft. Eng., vol. SE-23, n°3, pp 386-397.

Peter, T. (1990). Sûreté de fonctionnement: Modelisation par chaîne de Markov d'un système decrit par grafcet. D.E.A Automatique. Productique et theorie des systèmes .

Zhou,M. and Di Cesare, F.(1989) Adaptive design of Petri net controllers for error recovery in automated manufacturing systems. Rensselaer Polytechnic Institute, Troy, New York, 31p, .

This work is supported by the M.R.T. project (Board of Research and technology), with the collaboration of CETIM and SNPE.

A METHOD FOR CONSIDERING SAFETY
AND RELIABILITY IN AUTOMATION DESIGN

M. Reunanen and J. Heikkilä

*Technical Research Centre of Finland, Safety Engineering Laboratory, P.O. Box 656,
SF-33101 Tampere, Finland*

Abstract. In order to produce safe and reliable products at reasonable cost designers are advised to identify the fundamental weaknesses during the earliest possible design phase. The complexity of the functions that automated machine systems are required to fulfill, however, makes it difficult to follow this recommendation. An overall model that clearly specifies the system functions is needed. From the conceptual design phase the designer knows what functions must be fulfilled. By introducing appropriate tools the system's function structure can be established and used in checking the functions and their relationships.

The paper presents the application of Structured Analysis and Structured Design method for Real Time Systems (SA/SD) in specifying and modeling the function structure so that safety and reliability aspects can be systematically considered. The analysis of the established structure performed by using Hazard and Operability Study (HAZOP) is also presented. The method described has been applied to an intelligent gripper of a robot.

Keywords. Safety; reliability; automation; specification languages; describing functions; system analysis.

INTRODUCTION

The use of automation has grown rapidly with the welcome result that both productivity and worker safety can be increased. One of the means to reduce accidents is to introduce automation to those areas of work which are dangerous. However, many areas that actually need automation are left untouched because automation is not considered profitable. According to a Japanese investigation (Anon.,1985) mechatronics has been successful in raising productivity, enhancing product quality, and reducing labour needs, but, despite much talk of eliminating dirty and dangerous work, mechatronic equipment has not generally been installed for that purpose or at least, when installed, has not had the desired effect. Therefore, it is not self evident that the increasing use of automation will decrease safety problems.

Automation is often thought to introduce new hazards one of the reasons for this being the complexity of the functions that automatic machine systems are required to fulfill at high speeds, in large movement areas, with heavy loads, and with other energy potential present. Accidents have shown that, due to the complexity, it may be difficult for the operators, and even for the designers, of a machine system to adequately understand the functions performed by the system. Therefore, to cope with such complexity, methods are needed that can efficiently support the design and representation of the behavior of the system and which allow safety analyses to be made on the description.

We have adopted VDI Guideline 2221 (1987) as a model of the design process. According to the guideline, the conceptual design phase in the field of precision engineering consists of three stages: 1. Clarifying and defining the task, 2. defining the functions and their structures, 3. Searching for the solution principles and their combinations.

Accordingly, the results of these design stages are specification (requirements list), function structure and principle solutions for electromechanics, eletronics and software. This paper concentrates on the first and second of the above mentioned design stages.

PROPOSED APPROACH

In order to produce safe and reliable products at reasonable cost designers are advised to identify the fundamental weaknesses at as early a stage as possible. From the conceptual design phase the designer knows what functions have to be fulfilled by the product. Combination of the functions into a structure which presents the flows of information, energy and material, and their transformations, forms an early stage model of the product. VDI 2221 recommends that this model is documented. We believe that, if the model is properly established and documented, safety and reliability analysis methods may then be applied to the model to produce useful results.

Structured Analysis and Structured Design for Real Time Systems (SA/SD-method)

The function structure of a system is a model of what needs to be done by the system which is to be designed. One of the objectives of establishing a function structure is to enhance the basic quality of design, including safety and reliability, so that premature implementation can be avoided. SA/SD is one of the methods by which the function structure of a system can be modeled.

The properties of SA/SD and other modeling methods, e.g. SADT, Bond Graphs, Petri Nets and a method represented by Pahl and Beitz (1984) have been studied by Heikkilä (1989). SA/SD was considered to be the most appropriate for modeling systems in which all the basic flows (material, energy, information) and their transformations play an important part. The SA/SD-model is hierarchical and different types of the representation are available, e.g. a data flow diagram and a state transition diagram. A comprehensive description of the method has been presented by Ward and Mellor (1985).

Hazard and Operability Study (HAZOP)

The HAZOP study has been developed to examine material flows and processes taking place in pipelines and vessels. The purpose of HAZOP is to identify hazards and operability problems by searching for deviations from the way the system is expected to operate. In the search for deviations, a fixed set of words, called guide words, is applied to the system variables. The method is systematic and e.g. flow charts are used as a basis of the study. The details of the method can be obtained from the reference (Guide... 1977).

The basic idea of the proposed approach is that an SA/SD-description could be analysed by applying the HAZOP-method, thus making an early identification of hazards possible in a systematic fashion.

Strategy for Safety Engineering

According to the European draft standard CEN/TC114N93E (1988) all the hazardous situations in the various states of a machine, and relating to the various aspects of the man-machine relationships, should be identified. When every hazardous situation has been identified, the standard states that a following "three-stage method" should be implemented:

a) Avoiding the hazard or limiting the risk as much as possible by design

b) Safeguarding against hazards which could not be avoided or sufficiently limited on step a)

c) Informing and warning users about the residual hazards.

CASE STUDY

Application Object

The proposed approach was applied in a close co-operation with a design project whose objective was to design an intelligent robot gripper for handling fragile workpieces. The following operational requirements were drawn up by the gripper designers (Airila and others, 1989):

- the gripper must be able to handle cylindrical or spherical workpieces, the diameter range from 50 to 150 millimeters
   the gripper must allow machining operations, e.g. grinding or deburring
- fingers must be exchangeable
- the gripper has to be able to pick pieces from narrow spaces (workpieces situated side by side on a pallet)

the gripper must not weigh more than 2 kilograms
the gripper must be able to handel pieces of 0,05 to 1,5 kilograms. Acceleration of 10 m/s^2 and external machining force of 10 N are present
the gripping force per finger must be controlled on the range from 10 to 60 N
the workpiece is fragile and its surface is sensitive to contact forces
- pneumatic and electrical energy is available.

Designing for Safety and Reliability

The gripper design was started by establishing an environmental model of the gripper. The model consists of a context diagram and of an event list. The context diagram is presented in figure 1.

The context diagram includes a description of the interfaces between the system and the environment. Based on the context diagram, a list of events which may occur in the environment, and to which the gripper must respond, was drawn up. An example of the event list is presented in figure 2.

The potential accidents in the event list which were to be taken into account at the subsequent stages of design were identified by applying the principles of the Preliminary Hazard Analysis method. The objective was to identify the forms of energy which, when unintentionally released, may injure the operator.

The flows of the diagrams describing the system were specified more precisely in textual form in a so called data dictionary. The environmental model and the data dictionary replace the traditional requirements list.

The next step comprises a specification of the behavior of the system. Determining of functions and their structures (construction of the behavioral model) was started by drafting an overall flow diagram of the gripper. The draft included the relevant functions of the system. The highest hierarchy level of the model, the main function level presented in figure 3, resulted from the grouping of the functions of the draft model. The function synthesis progressed so that the main functions were further divided into subfunctions and described more precisely by the state transition diagrams. This is the latest point of the function synthesis at which the events written in the event list can be taken into account. For example, when establishing the state transition diagram of the function EXECUTE GRIPPING (figure 4) the following events have been considered: "gripping succeeds", "gripping is successfully released", "compressed air is cut off", "electricity is cut off", "compressed air is turned on", "electricity is turned on". When e.g. the compressed air or electricity is cut off the gripper is designed either to maintain its gripping force if a work piece is carried by the gripper or to keep the aperture constant if the gripper is empty.

The identification of the functional safety and reliability problems was carried out by applying the HAZOP-method. In this case HAZOP was used to check the effects of the deviations connected to the flows of the description. All the flows in all possible states of the gripper were not studied. Figure 5 presents an example of the HAZOP study. The example deals with the effects of the deviations in the flows from the environment. The gripper is in its automatically controlled state with a constant preliminary aperture.

The proposed method for considering safety and reliability aspects was developed by combining SA/SD, PHA and HAZOP methods. In the case study, the functional model of the intelligent gripper was created and analyzed with this method. The functional model created consists of 20 diagrams arranged in 5 levels of hierarchy, 15 transformations (basic functions), and 86 flows. Seven functional states of the system were modeled.

As a result of the HAZOP study, 17 suggestions were made to improve safety and reliability. The study covered about one third of the model (40 deviations of 27 flows) in one functional state. Twelve of the deviations were identified to have the potential to cause an accident, 2 of the deviations caused (in the state studied) only an alarm to be sent to the main system (robot), 16 of the deviations did not affect the function of the gripper, and 3 of the deviations were found to be nothing but the normal state of the flow (in the studied state of the system). In 7 cases it was impossible, or at least irrelevant at that stage, to define the consequences because of their obvious dependence on implementation technology. Suggestions to take them into account at the subsequent stages of the design were noted. The HAZOP study took about 8 hours of team time.

CONCLUSIONS

The purpose was to find or develop a method to improve functional safety and reliability of automation by means of design. One objective was to move the application of systematic safety considerations to the earliest possible design phase.

It was found that in practice a functional model is rarely created. In spite of that, the function structure was selected as the basis for safety and reliability analysis because of its potential to enhance the understanding of the system behavior and communication amongst different parties involved with the design. Unfortunately, that led to a situation in which we had to construct the functional model with a design group as a part of the safety analysis (which was not the actual intention). The design group considered the function structure helpful in that it aided understanding and communication, and thus improved the total quality of design. However, as a part of a safety analysis only it was regarded as too time consuming. The difficulty of describing environmental effects (e.g. power in this case) throughout the model hierarchy was considered a weakness of SA/SD modeling.

The fact that 12 potential event chains leading from deviation to incident was found in the HAZOP study of the functional model, implies that the model can be used as it was intended. The functional consideration of the safety and reliability problems identified was found to be easy and effective (at the functional stage). The problem is that the exact technical implementation of the planned behaviour of a system is very hard (if not impossible) to accomplish. Therefore, safety and reliability considerations within function design diminish but do not remove the need for later safety and reliability analyses.

The analysis of the 3 deviations (in the case study) that, in fact, were part of the normal operation is a consequence of the difficulty of understanding and sufficiently describing the behaviour of an automatic system. In this case, the problem could have been solved by creating separate functional models for each state of the system.

The proposed method is useful to improve the understanding of the behaviour and to design for the abnormal operating states of functionally complicated systems. The resources needed for the analysis part of the method are comparable to those of a normal HAZOP study. Although this method is based on SA/SD-method, same principles can be applied with other function design methods and documents describing the function, too (e.g. operating instructions as well as program code can be analysed). SA/SD-model can also be used to combine different documents (e.g. operating instructions and program code) as a total functional model of the system.

Most of the time spent in the HAZOP study of the functional model was in manual tracking of event chains. It is quite obvious that this task (or at least the main part of it) would be easy to automate, and thus time can be saved, because the SA/SD-model is by nature close to a programming language. Even if the tracking system was automated, tracking should be done step by step under the direction of the analyst. That is the most effective way to improve the understanding of system (model) behaviour and it allows the user to check the operation of the tracking system. The user should also be allowed to modify the functional model during the study and thus ensure that the model will be appropriate (it is unlikely that meaningful results can be obtained from HAZOP study of an improper model). These ideas have been applied in the development of STARS software package (Poucet, 1991). STARS includes tools for analysing safety and reliability.

The need for further development of the proposed method depends on whether SA/SD or some other formal function design method will be more commonly used or if the evolution of design methods follows another course. In the former case, computer aids could be developed quite easily.

REFERENCES

Airila, M., T. Ropponen, J. Merilinna, and M.Kiiskinen (1989). An intelligent robot gripper for handling fragile workpieces. In, Sensors, Vision and Inspection. A One-Day Conference at AUTOMAN-5, Birmingham 9 May 1989.

Anon. (1985). Mechatronics: The policy ramifications. Asian Productivity Organization, Tokyo.

CEN/TC114N93E (1988). Safety of machinery. Basic concepts. General principles for design. European Committee for Standardization, Draft, 1st July 1988.

Guide... (1977). A guide to hazard and operability studies. Chemical Industries Association Limited and Tonbridge Printers Ltd., Tonbridge.

Heikkilä, J. (1989). Designing for safety and reliability of the Operation of a Technical System. M.Sc. Thesis (in Finnish), Tampere University of Technology, Tampere.

Pahl, G., and W. Beitz (1984). Engineering Design. The Design Council, London.

Poucet, A. (1991). STARS: Software tools for analysis of reliability and safety. Project description. In, STARS Project Conference hosted by VTT (Technical Research Centre of Finland), Tampere, June 5 - 6, 1991.

VDI 2221 (1987). Systematic Approach to the Design of Technical Systems and Products. VDI-Verlag GmbH, Düsseldorf.

Ward, P. T., and S. J. Mellor (1985). <u>Structured Development for Real-Time Systems.</u> Vol. 1 and 2. Yourdon Press, Englewood Cliffs, New Jersey.

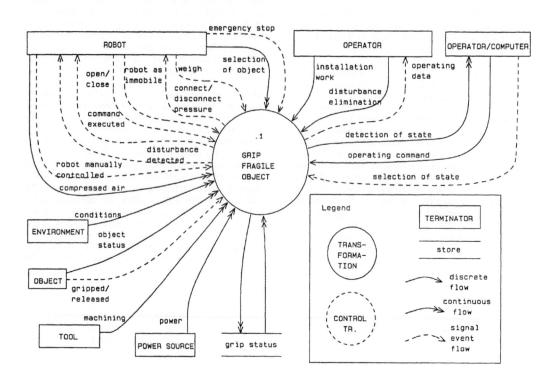

Figure 1. Context diagram of the gripper.

INTELLIGENT GRIPPER
M.R., J.M., S.H. & J.H.

13.2.1989
EVENT LIST

normal event	deviation	accident
ROBOT		
Commands to grip		
Commands to release	Faulty control command	Gripper grips robot
Selects the object	Faulty object selection	Gripper collides with robot
Turns compressed air on	The pressure of compressed air	
Turns compressed air off	increases or decreases over the	
	threshold values	
	The pressure of compressed air	
	returns to the allowed value	
	The flow of air deviates from the	
	allowed values	
	The flow of air returns to the allo-	
	wed value	
OPERATOR		
Begins to install	Faulty installation	Worker collides with gripper
Is Installing		Gripper collides with worker
Stops installing		Gripper grips worker
Begins to eliminate disturbance		Electricity injures worker
Is eliminating disturbance		Compressed air injures worker
Stops eliminating		

Figure 2. A partial event list of the gripper.

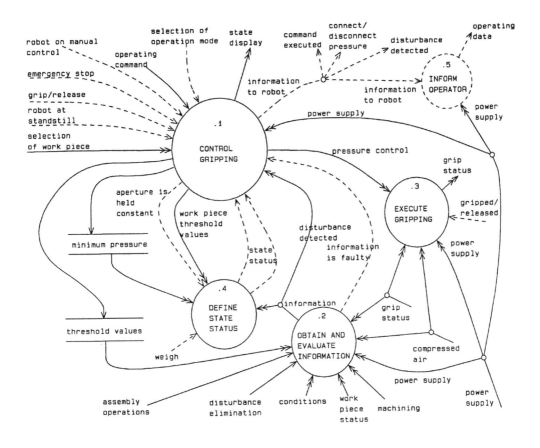

Figure 3. The main functions of the gripper.

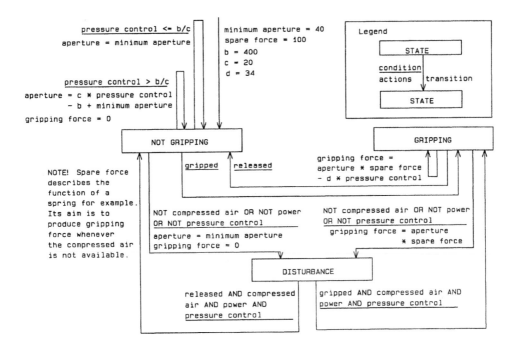

Figure 4. State transition diagram of the function EXECUTE GRIPPING.

Techincal Reasearh Centre of Finland Safety engineering laboratory HAZOP	SYSTEM: Gripper STATE: Automatic operation ACTIVITY: Moving TRANSFORMATION (FUNCTION): Grip fragile object	PAGE: 1 (2) DATE: 5.1.1989 AUTHORS: M.R., M.K. J.M., S.H. & J.H.

FLOW	STATE OF FLOW (deviation)	CONSEQUENCES	MEASURES
Compressed air	Low pressure	1. "Disturbance detected" signal to robot 2. Gripper to minimum aperture -> gripper collides with object	Function must be modified so, that aperture does not change. Robot ougth to stop the motion and if possible eliminate disturbance
	High pressure	A meaningful answer is not yet available.	Remember! (At the subsequent stages of the design)
	Low flow	Normal function	
Robot manually controlled	On	No effect	
	Off	No effect	
Close	Signal	1. Grip -> "Disturbance detected" signal to robot 2. Gripper returns to the preliminary aperture and (possibly) collides with object	Robot should stop the motion and command gripper to open. When "disturbance detected" signal has vanished, robot can go on working.

Figure 5. HAZOP record sheet.

112

AN EARLY WARNING METHOD FOR SAFETY-CRITICAL SOFTWARE DEVELOPMENT

A. D. Pengelly

BT Laboratories, Martlesham Heath, Ipswich, Suffolk IP5 7RE, UK

Abstract :
Measurement and specification are important aspects of the 'classical' engineering disciplines, so why not software engineering? This question is particularly pertinent when considering the development of safety-critical systems. This paper describes a technique which, via the use of specification, measurement and basic statistics, can provide the project manager with a useful aid for controlling such complex and costly developments. The concept of an *early warning method* is proposed.

keywords : specification, measurement, cost, complexity, management.

1 Introduction

The use of software based systems for telecommunications applications has become common practice. Applications include digital exchanges, automatic fault location and network management systems. All rely on software in potentially high risk [1] areas and there is sufficient evidence to suggest that the use of software in these situations is expanding rapidly.

The various application domains, and the software subsequently developed, are becoming increasingly more intricate and expensive. Also the demands for higher levels of reliability, maintainability and reduced costs (among other things) further exacerbates an already difficult problem.

The development of such software is a most complex task. It can also be a black-hole regarding resources, resulting in man-years being consumed with apparently little return on investment. As a result these projects can quickly become very difficult to control, both from a technical and managerial perspective.

The key problem is the sheer complexity of the system being developed. Hence the central issue is one of *complexity and cost management*, and the development of appropriate procedures and methods to enhance the project managers capabilities in these areas. What is required is a technique or method which can provide the project manager with a means of extracting pertinent information from the abstract system representations, early in the life-cycle. This paper develops the concept of an *early warning method* for the manager involved in the development of high-risk software. The analogy with the military early warning systems is quite deliberate. The project manager needs to develop a strategy for coping with the complexity of the product, and to prepare adequate defences and hence keep costs down to an acceptable level.

The aim is to provide, via measurement and specification, a means by which the manager can identify anomalous components of a system at a very early stage in the development life-cycle, thus providing them with the facility to allocate resources effectively and to develop appropriate contingency plans. The key objective is relative simplicity avoiding the , perhaps natural, trap of developing

---

[1] The risk is predominantly financial.

a complex solution for a complex problem. This objective is achieved by using established software measures in conjunction with relatively simple statistics.

The paper will discuss the basic procedures and methods required to develop such a method. The key ingredients are a formal means of system representation and an associated measurement set. It will be shown that these two, rather distinct, areas of software enabling technology can be made to work in a symbiotic fashion, and provide the project manager with a useful control aid.

2 An Early Warning Method

The project managers view of the situation can thus far be summarised as follows :-

- safety-critical systems can be difficult to manage and control

- formal methods are not widely used because they are difficult, time consuming and appear to present little in the way of return on investment

- measurement is generally seen as a good idea, but demonstrable benefits are hard to come by

The early warning method currently proposed is but a small step in combining the principles of measurement and specification in order to facilitate the management of such projects. More advanced applications are currently under consideration.

The rationale behind the method is to measure key characteristics of the system specification, and via the use of basic statistics attempt to identify potential problem components early in the life-cycle. This technique, referred to as outlier analysis, is generally available on most statistics packages.

It should be stated that without the appropriate tool support the method described would be impractical. Hence there is a need for somewhat specialised tools to be developed.

BT has been involved in the development of two such tools. One as part of an Esprit II project, the other internally. The former is

the Esprit II project COSMOS (COSt Metrics Of Specification) [3, 5], which is based on Fenton-Whitty theory [6]. The latter a design metrics tool based on work carried out by BT in collaboration with Shepperd [2] and Ince [3] [8].

The platform adopted for the experiment described in section 3 was the COSMOS tool. The reason for choosing this tool is that it is the only CASE tool known to the author, capable of automatically measuring formal specifications (principally LOTOS [1] and SDL [2]).

Fenton-Whitty theory [6] was initially devised as a graph theoretic approach to the mathematical modelling of structured programming, enabling objective measurement of software using structural characteristics. The assertion is that a program consists of a collection of basic constructs called *primes* [4] which, via the operations of nesting and sequencing, completely specify the abstract system representation. Whilst the theory was initially developed for imperative languages, it has been extended to cover formal notations such as LOTOS [4]. It should be pointed out that the theory is unlikely to be applicable to all formal languages. In fact languages such as **Z** for instance do not seem amenable to prime decomposition. However it is felt that some notion of control-flow is discernible within LOTOS, and as a result the application of Fenton-Whitty theory is not entirely inappropriate, though there are research issues that need addressing [4].

3 Analysis and Interpretation : A Case Study

In this section the analysis of a LOTOS specification is discussed. The idea has been to adopt a pattern of reasoning which is likely to be used in practice. Hence complex statistics and inference analysis is avoided. The approach can be described as intuitive but based on objective, though perhaps incomplete, information.

This section presents a typical scenario of

[2]Bournemouth Polytechnic
[3]Open University
[4]This definition is not to be confused with that of standard graph theory.

how the ideas discussed could be used in practice. The data presented is from a real project. In this example the COSMOS tool is being used as an 'early warning method'.

For commercial reasons the system being specified cannot be discussed, apart from the fact that it relates to a telecommunications system. However, as the reader will see, this is unimportant from the point of view of the analysis.

3.1 A Possible Scenario

Consider figure 1. It will be seen that, along with a clear correlation between two metrics, there are two points (LOTOS processes) which do not conform to the anticipated behaviour. The metric '# primes' refers to the number of prime constructs within the process structure, whilst '# nodes' is the number of process nodes. Concentrating on one point, what can be said about the process at point a? Whilst being a 'large' system, for some reason it is not being decomposed according to the theory.

The first interpretation is that the structure of the process is unusual in some way, that the basic constructs (primes) do not exist or are not discernible. This would seem to imply that the process is not well structured in the usual sense of the word. Analogous phenomena can be observed in imperative languages when **goto** statements have been used.

The point is that now the project manager is *aware* that one aspect of the system is exhibiting unusual behaviour, that it is an outlier. There is at this stage no requirement to know in detail what the process does or the technical details associated with LOTOS. The information presented is qualitative. It is merely 'ringing an alarm bell', warning the manager of a potential problem. Of course the first thing the manager may do is to set up a review meeting to discuss the *problem* in more detail. They may wish to use the data for risk analysis purposes (BT has developed such a procedure [7]). Here the impact and consequences of the anomalous data could be catered for by deriving appropriate contingency plans. They may also wish to look at other metrics in order to acquire more information. Even if they

do nothing but take note of the data, a potential problem has been highlighted which might not otherwise had been anticipated.

Figures 2 and 3 presents alternative views of the same specification. Comparing the number of nodes with two complexity measures [5] again reveals the same outliers. This data seems to add credence to the belief that the two components of the system, identified using figure 1, are indeed somewhat unusual.

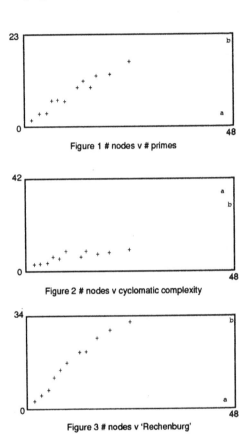

Figure 1 # nodes v # primes

Figure 2 # nodes v cyclomatic complexity

Figure 3 # nodes v 'Rechenburg'

Figure 1:

It may be that the process needs to possess these properties in order to fulfill its function. It may be that perhaps a new recruit wrote the specification and is struggling with the notation. Nevertheless contingency plans can now be drawn up. In addition the manager is less likely to encounter any surprises in terms of such components being discovered later in the life-cycle, disrupting

[5]McCabe is the well known cyclomatic complexity metric. Rechenburg is a, as yet unpublished, metric developed by Alcatel.

plans and resource allocation. The example shown is from a real project, and indeed the process concerned was abnormal and difficult to derive. It should be stated that the team responsible for this particular project felt that the process in question could not have been changed significantly. However they also pointed out that this process only gradually emerged as a problem, and that it would have been useful to identify it earlier.

The concept has also been applied to complex data processing systems, using the design metrics developed by BT along with Ince and Shepperd, with encouraging results. The emphasis of this work has been targeted at the systems intrinsic data structures and information flow [9]. This work shows that the underlying syntax and semantics need not be that 'formal' [6].

Larger experimental field trials are currently under preparation.

4 Conclusions

The *concept* of an early warning system has certain attractions to project managers and analysts. Accepting that only broad indicators can be extracted from the abstract system representation by the methods discussed in this paper, the project manager or analyst can adopt other techniques in order to home in on the problem. In the short term there is likely to be a trade-off between the detail or accuracy of the information obtained and the stage in the life-cycle from which it was extracted. Hence information obtained early in the life-cycle is likely to be somewhat coarser than that obtained later. However from the point of view of a 'warning system', this is acceptable. That the system is able to flag possible future problems is in itself very useful information at a very important point of the systems evolution.

References

[1] ISO, Information Processing Systems, Definition of the Temporal Ordering Specification Language, *TC97/16 N1987, 1984.*

[2] Faeremand, O et al. SDL'89 : The Language at Work. North-Holland, 1989.

[3] COSMOS (Esprit II Project 2686), Technical Annex, December 1988.

[4] Pengelly, A D and Fuchs N. Software Structure and Cost Management. *British Telecom Technology Journal, April 1991.*

[5] Pengelly, A D and Simons, C A. A Front-End Measurement Tool based on Fenton-Whitty Theory.*Proceedings 3rd Annual Oregon Software Metrics Workshop, 1991.*

[6] Fenton, N E and Whitty, R W, Axiomatic Approach to Software *Proceedings 1st European Seminar on Software Quality, Brussels, 1988.*

[7] Steele, S. SERAM. *Proceedings European Software Cost Modelling Meeting, 1991.*

[8] Ince, D. Shepperd, M. Pengelly, A and Benwood, H. The Measurement of Data Designs. *Proceedings 3rd Annual Oregon Software Metrics Workshop, 1991.*

[9] Ince, D. Jackson, M. Pengelly, A and Shepperd, M. Designing Data Models using Metrics. *Wolverhampton Polytechnic, School of Computing Technical Report, 1990.*

---

[6]This begs the question of what level of formality is required in the notation

A CONFIGURATION MANAGEMENT SYSTEM FOR INCREMENTAL DELIVERY PROJECTS

F. Redmill

Consultant in Management & Software Quality, 22 Onslow Gardens, London N10 3JU, UK

ABSTRACT

In all software development projects, configuration management is crutial and requires careful planning and control. In Incremental Delivery projects, it is even more important and more complex, as a number of versions of the system are simultaneously in existence – one in operation and at least two in development. This paper reports on the author's experience with configuration management in Incremental Delivery projects. It explains the method and procedures developed to control the software, how a proprietory configuration management system was tailored to support the method, and how the system was managed.

KEYWORDS: Incremental Delivery, Configuration Management, Software Development.

INTRODUCTION

Software configuration management concerns the control of software at all stages of its life cycle (e.g., unit, sub-system, integrated system, released system) in order to minimise errors, facilitate access to and correction of the software, and ensure, among other things, that:

(i) Testing is methodical and thorough;

(ii) Every unit is always traceable;

(iii) All changes are controlled;

(iv) Systems are built only of the units and subsystems which have been tested together;

(v) Documentation for all levels of development exists and is accurate and up-to-date.

In any development project, software configuration management is important. In an Incremental Delivery project, it is doubly so; and it is more complex (REDMILL 1989). Even before the first delivery is made, the second delivery is planned and its development is underway. The composition of each delivery therefore needs to be defined, and its software isolated from that of all others.

With the first delivery, the system becomes operational. At the same time, development must continue, so there has to be more than one version of the system in existence at any given time, one live and one or more under development. As each delivery must go through the Waterfall life cycle of planning, specification, design, build and module test, integration test, and system test, it is not feasible to have the entire development team working simultaneously on one delivery, and having only one version under development at any given time is an inefficient use of staff skills. Specific disadvantages of doing this are:

(i) Staff are not used optimally;

(ii) There is a danger that too much new software is included in a delivery, and this increases both the difficulty and the duration of debugging;

(iii) If delivery N is not commenced until delivery N-1 has been delivered, adhering to the agreed periodicity becomes more difficult, and giving the customers and users early information on the content of a delivery becomes precarious.

It is therefore important to define both a method of configuration management and a staff organisation which allows the concurrent development of more than one delivery, optimizes the use of staff, and allows accurate planning of the content and dates of deliveries. This paper describes the configuration management support system which was developed by the author's team in British Telecom, the facilities it offered, and the way in which it was used and managed.

THE NEED FOR A DEVELOPMENT PROCEDURE

There are many configuration management tools to aid developers in controlling their software. However, like all software tools, they are hardware dependent - or, at least, system-software dependent. Thus, if you standardize on a tool before selecting your development system, it is likely that you will have to use your tool off-line. This has the disadvantage that it requires great discipline on the part of the developers to record every change to the software at the time it takes place.

In our development, we used a software configuration management system (CMS) which was integrated into the total development system, and which provided a software library within which software at all stages of development was stored. It therefore possessed the potential to provide configuration management automatically, thus relieving the developers of the responsibility and the overhead of separate configuration management activities. For this potential to be revealed, however, it was necessary to define the development process, step by step, and then to tailor the CMS to facilitate the process. In the case of configuration management, it is necessary to take time to do the following:

(i) Define the development procedure;

(ii) Document it;

(iii) Define a procedure for

controlling the documentation, with minimum requirements of label, date, author, document number, and issue number, and issue it to all staff;

(iv) Ensure that all staff know and understand the procedure;

(v) Tailor the proprietory CMS to support the procedure;

(vi) Tailor the proprietory CMS, as far as possible, to ensure that only that procedure can be used, or, at least, to make any other method of working unattractive;

(vii) Develop any software tools necessary for implementing the previous two items.

At the top of the list is the working procedure for the development process. Everything else is built around this. However, while it would be nice for the Development Manager to devise an ideal method of working and then purchase a configuration management tool to support it, it is unlikely that the right tool would be found. It is more likely that the working procedure will depend, to some extent, on the capabilities of the support tool. Thus, the first step is to consider the hardware and system software which have been chosen for the development system, and to select a configuration management support tool which is compatible with them. Then the development procedure must be defined, publicised, understood, universally accepted, and adhered to. There is no single right way of working, but the next section describes the method which we designed and found to work well.

THE CONFIGURATION MANAGEMENT SYSTEM

Fundamentals

We purchased a CMS which was compatible with the hardware and system software of our development system. It was integrated into the system to provide a library within which to store and control the software. However, it was up to us to decide how to employ this library to achieve this control at all stages of the software's development and maintenance. The library, in its initial state, allowed only one version of the software to be stored at any time. The first thing we did was to partition the library so as to define five levels of software, and this, in effect tricked the CMS into treating each level as a library. We were able to store a version of the software system at each level and, thus, to keep a number of versions under development concurrently.

We also provided a working area at each level, partitioned from the system storage area so that changes made in the working area would only be integrated into the system intentionally and after thorough testing. The levels of the library (and therefore of our software) are shown in Figure 1 and are:

> Level 1,
> The Test (T) level;
>
> Level 2,
> The Integration (I) level;
>
> Level 3,
> The System (S) level;
>
> Level 4,
> The User (U) level;
>
> Level 5,
> The Live (L) level.

As suggested in the discussion of principles in the previous section, the mode of software development is governed by the facilities provided by the CMS. Thus, our design of the library was, first, a result of our planning of development management, and second, both a tool and an imposer of discipline on the development process. Tools were then developed by us to move software units and systems upwards from one level to the next. Two rules were made, and the tools were designed to support them:

(i) No downward movement is allowed;

(ii) No skipping of levels is allowed.

The first rule ensures that configuration records are not corrupted by indiscriminate changes. (However, changes made in the course of corrective maintenance need to be reflected in the lower levels, but the mechanism for achieving this is not the subject of this paper.) The second rule ensures that no item of software advances until it has passed all the appropriate tests.

Software Progress

Each level of the library possesses its own database which is tailored to the needs of storing and testing the software at that level. However, initial software development is carried out outside the library. Figure 2 shows that an individual programmer constructs a module of software to a specification, designs tests for it, and uses his own test database for testing the unit until he is convinced of its quality. He then transfers it to the T level (the lowest level) of the library, where it is tested by a member of the independent verification and validation team, using test specifications and test data previously prepared by the module's designer. Test reports are prepared and copied to the Development Manager and the programmer, and time spent by the latter on the rework of faulty software units is recorded as a quality-related cost, as is the time taken for re-testing.

A unit of software does not have a standard, or even a maximum, length. With experience, we found it most practical to commission a programmer to build a function of the system (rather than merely the lowest level of module), while encouraging him to decompose this into as many modules as possible. Thus a programmer coded, compiled and tested small modules, and then integrated them into a function which he tested before introducing it into the T level of the library. On one particular project, the average length of functions written in a 4GL was about 2000 lines of code, and these were mostly transaction processes. 3GL functions averaged 300-400 lines of code, and many of these were interface communication functions.

When the unit has been proved to conform to its specification at the T level, it is passed to the I level, where it is integrated with other units of software, written by the same programmer and others. The integration is carried out according to design plans, and each integrated sub-system is tested by the verification and validation team, using test specifications and test data prepared earlier by the designers. Again, test reports are compiled, filed under the reference of the delivery in preparation, and copied to the Development Manager. The programmers are informed of the test results and, again, rework and re-testing at the T and I levels, due to programmer error, are recorded as quality-related costs.

At these two levels, where it is hoped that most program errors will be detected and corrected, the modules and sub-systems are compiled with a debugger. This creates an overhead of greater compilation time, but it provides the facility to step through the code, instruction by instruction, in search of an elusive bug. The debugger is seldom needed at the next three (system) levels, and is not used in the first instance. It is called into play, however, at the expense of considerable processor time, when a serious intermittent bug is present.

Successive levels of integration

testing are carried out until, finally, the whole system for the delivery in preparation has been tested. This is then passed to the S level.

Until now, tests have been designed to find bugs in the system and to verify that the software units conform to their design. Now, at the S level, validation tests are carried out to prove the functionality of the system as a whole and, thus, that the right system has been built. In other words, the product is not being tested for accuracy in the translation from a preceding stage of development, but for conformance to its original specification of requirements.

When effectiveness has been proved, the system is moved to the U level where it is made available to the customer for pre-delivery testing. It is stored there until it is time to prepare it for delivery, when it is passed to the L level. Depending on the allocation of the customer's staff to the project, they may or may not choose to avail themselves of the opportunity to carry out pre-delivery testing.

Prior to the delivery date, the system is passed from the U to the L level. On one project, where the system grew to almost a million lines of code, the time before delivery when this needed to happen was between one and two weeks. During this time, the code for the units of the system is generated, and the system is built and compiled (the method of storage of units is by original version plus successive changes, see the next section below). Then, each module used is recorded, with its version number and with details of its links to other units, so that a complete configuration profile is created for the delivery. This is then documented and stored. Next, a number of confidence tests are carried out to confirm functionality, check the new features, and ensure that the

system is that which was validated at the S level. Finally, at the appointed date and time, the system is delivered to site. In the early days, we loaded the system onto magnetic tape for transportation, but latterly we set up a direct link to each system in the field. Over these links we not only delivered software, but also controlled the systems when necessary for maintenance.

Version Control and Storage

At each level of the library, there is a control program which records the presence of every unit of software which arrives at that level. For a new arrival, it records its name and gives it a version number; for a unit of the same name as one already encountered, it compares the new code with the old. If there are changes, it records them and allocates an up-dated number to the new version; if there are none, it does not change the version number. Having thus attended to identification, the control program must now organise the storage of the unit. What it actually stores are the original version and the successive changes made to it, each with its version number. This has the advantages that storage space is saved and that previous versions are always readily obtainable, which is helpful when a new version causes serious problems.

However, there is also a serious disadvantage. This is that, at each level, the software must be both rebuilt and recompiled before it can be tested. At the T and I levels, this is not too much of an overhead, but it certainly is at the S, U, and L levels where the new version of the complete system is under consideration. A resulting further disadvantage is that, following system testing at the S level, the system is rebuilt twice more (at the U and L levels) before being delivered to site. A great deal of care therefore needs to be taken to ensure that the system built and compiled at the L level is exactly that which was tested at the S level. We thus developed a program to check this. It makes a record of the modules included in the S-Level system, with their version numbers. It is then passed up with the system. At the L level, it is activated to test the newly built system against its own stored record, and it lists and prints out discrepancies. Of course, this program is stored in the same way as other modules, and it is not delivered to site as part of the system.

This method of storage was designed to save space. With a system as large as ours, it would not have been possible to store successive applications in their entirety. As shown above, the penalty was a time overhead. Moreover, as the number of versions increased, so did the time overhead, as each version could only be constructed from Version 1. We therefore decided to rationalise this by eliminating the earliest versions. Study of the pattern of our access to earlier versions led us to conclude that we were safe in retaining only four previous versions — which represented a year in time. Recreating the current system therefore involved applying four sets of changes to the stored version, instead of a number which increased with time.

MAKING CHANGES

A bug may be detected in the software in any of the five levels of the library. While it is our policy for the defective module to be corrected at the lowest level of its existence, there may be a need to access a module at any level and, sometimes, to make an immediate correction. For example, if an application at the S level is faulty, but each of its component modules successfully passed their I- and T-level tests, it is necessary to find and correct the

fault at the S level before reintroducing the changed modules at the T level, testing them, and passing them up through the levels. Thus, we made it possible to access modules at all levels, via the working areas already mentioned and shown in Figure 1. At the same time, however, we recognised the need to guard against two programmers making concurrent changes to a module. (Then the first new version to be replaced in the library would be overwritten by the second.) We thus allowed access to the software by the use of two commands.

The FETCH command allowed a copy of a module to be taken but not replaced. A programmer could thus take a copy into a working area of storage, but any changes made could not be introduced into the library.

The RESERVE command allowed a copy of a module to be taken but, when this was done, it required the programmer's name and password to be inserted. It then validated that programmer's right to make changes to the software and "reserved" that module, at that level, for that programmer. No one else was then able to RESERVE the module until the programmer had either replaced it, with the same or a changed version, or cancelled its RESERVE status.

Changes to the system were thereby possible under any circumstances, including emergencies, but they were always controlled. However, control must include recording the changes in a way which allows easy access to earlier versions and historical documentation to be maintained. Because a change may be made at any level, a simple version number which accompanies a software unit as it is passed up the levels would not guarantee uniqueness of change identification. For this reason, a unit does not take a version number with it as it goes from one level to another. Instead, it acquires a "Generation Number" at each level.

When it enters a level for the first time, it is allocated Generation Number 1 and, each time it is changed at that level, either becuase of a change made at that level or because it is passed up with a change, the Generation Number is incremented.

Thus, a software unit has a unique identity at each level. If its history is required, the Generation Numbers at each level, along with the datess and times of their creation, are printed out and put into chronological order.

MANAGING THE CONFIGURATION MANAGEMENT SYSTEM

Each level of the CMS library is the responsibility of a named manager. The structure of the development team is shown in Figure 3, and responsibilities for the levels were allocated to the managers whose roles were appropriate to them. Thus, the L level was under the jurisdiction of the System and Application Support Team Leader, the U and S levels were under The System Test Team Leader, and the I and T levels under the Low-level Design and Coding Team Leader.

The main responsibilities for each level are maintaining the integrity of the software within the level and ensuring that it is not advanced to the next level until it has passed all appropriate tests and the manager of the next level is ready to receive it. There has always been debate as to the best means of authorizing transfers from one level to the next. Leaving the manager of a lower level to pass software up when it was ready rather than when the manager of the higher level was ready to receive it was found to be risky. In most cases, all went well, but, when a problem occurred, there was likely to be a disagreement. Our solution was to lay down a procedure which decreed the following:

(i) A transfer should only take place when the manager of the higher level was ready to receive the software;

(ii) The manager of the lower level would document a list of all the modules to be transferred, and their version numbers, along with a statement that they had been thoroughly and successfully tested according to the predesigned test specification and test cases;

(iii) The manager at the higher level should check the list, ensure that all, and only those, modules which were expected (i.e., necessary for the delivery in question) were included, and then agree to accept the transfer.

Under this procedure, any discrepancies between what was expected and what was offered could be resolved before the transfer was made. Reference to the plan for the delivery was almost invariably sufficient to reveal which modules were missing or surplus. The content of modules did not need to be checked at this stage. If a module was present and of the correct version number, it was fair to assume that the changes made to it had been those specified. If they weren't, responsibility was in any case clearly identified. We found that this procedure and the clearly defined management responsibility for the integrity of the levels were sufficient to ensure harmonious, accurate, and timely transfer of the software up the levels.

CONCLUSIONS

This paper has described a configuration management system designed for use in Incremental Delivery projects where there are a number of versions of the software system in existence simultaneously.

It was shown how essential it is to design a procedure for controlling the software throughout its life cycle, rather than simply relying on a proprietory configuration management system without understanding its method. The system is a tool and, if a tool does not support a method, it is likely to be more of a hinderance than a help.

The control procedure designed for the author's projects was described, and it was explained how the proprietory configuration management system was tailored to support it. The manner of the system's operation was also explained, and it was shown how the software progresses through the system as the development process is carried out. How the software is stored, and how changes are made to it were emphasised. Finally, it was explained how the development process and the configuration management system are managed.

It was seen that control of software in an Incremental Delivery project is much more complex than it is in a Waterfall Model project, and that it requires a great deal more time, care and management.

ACKNOWLEDGMENTS

The author acknowledges the work and collaboration of his colleagues in British Telecom (NIS) which led to the development system described in this paper.

REFERENCES

Redmill FJ (1989). Computer System Development: Problems Experienced in the Use of Incremental Delivery. SAFECOMP 1989, Vienna, Austria.

FAULT AVOIDANCE THROUGH A DEVELOPMENT ENVIRONMENT ADOPTING PROTOTYPING[1]

S. De Panfilis

Engineering - Ingegneria Informatica S.p.A. Laboratorio di Ricerca e Sviluppo, via del Mare 85, 00040 Pomezia, Italy

Abstract.It is widely accepted that prototyping is a powerful technique to carry out early fault detection. Main features of prototyping are: feasibility check against project constraints; check of the designer's interpretation against user requirements; verification of the user requirements against his/her real needs. In spite of these advantages prototyping is not widely used because of some weakness of the current prototyping tools: weak connections with the system high level model; lack of connections with the target system code generator. This paper describes a software development environment, called DEEnv (Design Enhanced Environment), giving a partial answer to these two problems and carrying prototyping approach towards fault avoidance. DEEnv is an environment which automatically produce the prototype as a consequence of the high level design. This prototype can be used to refine, using an iterative and incremental method, the high level design minimizing in such a way possible design faults. The designer interacts with DEEnv using graphic tools whereas the code which describe the system model is generated; the system model is wholly described by an executable modelling language (GALILEO).The paper presents the main advantages and problems of the prototyping techniques. Then it analyzes the architecture of DEEnv and its impact on global quality of the software project.

Keywords.Software engineering; software tools; modeling; fault avoidance; prototyping.

INTRODUCTION

It is widely accepted that one crucial point in designing application is the communication among designers, users and commitment. Furthermore, it is strongly accepted the needs of environments to support software production so that their product is of high quality. To solve these needs several languages, techniques and technologies have been proposed; as a matter of fact the existence of methodologies and tools to support software production is currently a common presence in software engineering organizations.

A central role in an information system life-cycle, as it is described in (Boehm, 1976), is the one represented by Conceptual Modelling. The goal of this step is the production of a formal and complete description of the new system. This description, called the Conceptual Model, is given using a formalism which describes *what* the system does rather than *how* the system works.

The main improvements due to Conceptual Model

---

1 This work is partially supported by the National Council of Researches of Italy within the project "Sistemi Informatici e Calcolo Parallelo".

The main improvements due to Conceptual Model are:

— it allows the designer to represent the problem domain at the correct abstraction level;

— it is the basis for the implementation of the new system;

— it represents the document which is the basis for the maintenance of the system or to develop new integrated systems (Benedusi and others, 1990);

— it is the only system representation around which designers, end users and commitment meet together.

Several graphic formalisms exist aiming at suitably represent the Conceptual Model; in particular some of them are developed with the intent to solve the communication problem, e.g. *Data Flow Diagram* representation (DeMarco, 1978). It is correct to say that these formalisms represent an important improvement with respect to communication, but some problems still remain open. In spite of their simplicity the formalisms are not proper to users and to applications topics, so that it remains difficult to the users to understand if the model represents correctly his/her ideas or even if the new system will solve his/her Information System problems.

Furthermore, more than one formalism is used to build the Conceptual Model depending on the aspect being modelled; in general these formalisms are not integrated among them. This fact depends on the multiplicity of methodologies used during Conceptual Modelling and on the fact that these methodologies are not conceptually integrated among them. For instance, in general *Structure Analysis* (DeMarco, 1978) is used to model functions while the *Entity-Relationship Approach* (Chen, 1977), and its extensions, is used to model data.

The lack of a only one formalism to build the Conceptual Model not only complicates the model comprehension, but also does not allow the possibility to have a formalization able to support the execution of the model. As a matter of fact, an execution of the new system model, which allows the user to use the model itself as it will work on the new system, lets the user to really understand what will be the new system. So that the prototyping of the model seems to be a solution to check the

completeness and correctness of the Conceptual Model.

Prototyping here is intended an activity which supports a better comprehension of user requirements and which is completely integrated in Conceptual Modelling, because it is the model itself which is directly executed being expressed using a formal specification. A similar approach is presented in (Henderson, 1986), while a complete and detailed description of software life cycle paradigms supporting prototyping is exposed in (Agresti, 1986). It is important to notice that the Conceptual Model of the system, being the model expressed with an executable formalism, is executed without using the target environment to obtain the same goal, as it is proposed in other approaches, e.g. (Goodman, 1980).

With prototyping not only the user, but also the designer may really check the model both for correctness and for completeness. This way it is possible to say that prototyping represents not only a powerful means of communications, but also a thinking tool.

Obviously, this kind of reasoning it is not new in literature, but the actual lack of tools allowing prototyping during Conceptual Modelling, with particular regards to the design of Business Information Systems, has historic motivations based on the unsatisfactory outcome occurred by the first tools using this capability. These motivations are:

— *Prototype misunderstanding*: can be noticed a clearness lack on prototype goals between prototype realizers and prototype users; generally the end user thinks to have a complete new system in a *smaller environment*.

— *Prototype off-line production*: the prototype code does not derive automatically from the Conceptual Model; thus not only additional work is needed to produce prototype, but also the risk of not alignment between prototype and Conceptual Model arises.

— *Prototype unreusability*: the prototype code is not reusable in the following software development phases; thus, even clearly exist prototype use benefits, the prototype costs are not justified by the code reuse.

Currently tools supporting prototyping may be divided into the following categories:

— those which have as platform the target environment (e.g. TELON, TRANSFORM); these tools in general do not cover the Conceptual Modelling phase, but the subsequent phases as Detailed Design and Coding.

— those which use code generators which or stay or simulate the target environment (e.g. CorVision, GOLDRUN, IEF, Software trough Pictures).

Thus having in mind the Information System designers requirements and previous works (Auddino, 1991; Braegger, 1985) and toold, in this work it is presented DEEnv (Design Enhanced Environment) as an environment which supports Conceptual Modelling and which gives a solution to the previous questions: 1) uniform formalism to describe the Conceptual Model; 2) rapid prototyping of the model itself.

DEENV ARCHITECTURE

The main aspect of DEEnv is that it is a design environment based on the use of the Galileo language (Albano, 1985). Galileo is an executable language designed to suitably model Database applications. Main characteristics of Galileo are:

— Galileo is a conceptual language and supports the abstraction mechanisms of semantic data models: classification, aggregation and generalization (Hull and King, 1987);

— Galileo is a strongly typed language; every

Galileo expression has a type, which is determined statically. The Galileo type system guarantees that any expression that can be defined will not generate type errors at run time;

— Galileo has a rich set of data types and type constructors; moreover the user may define new abstract data types, which are indistinguishable from the Galileo's predefined ones;

— Galileo has type inheritance;

— Galileo is a statically scoped functional language;

— Galileo can adopt as its own predefined functions, functions which are written in foreign languages.

Thus, the property of DEEnv to be based on Galileo allows to give an answer to the problems exposed.

— uniform formalism: Galileo allows to describe completely all the aspects of the Conceptual Model of an Information System.

— rapid prototyping: Galileo is an executable language.

However, a lot of work and experience is required to build a Conceptual Model of an Information System using Galileo. Due to this reason and to some others, which are following explained, a computer environment which supports the designers in their work has been planned. This environment, DEEnv, is made up of the following main components (see

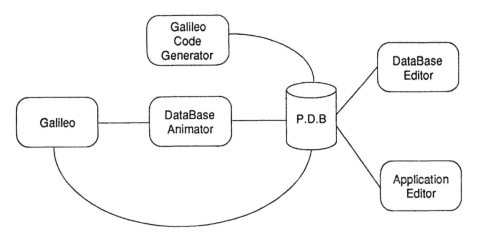

Fig. 1 - DEEnv architecture

Fig. 1):

— a graphic editor to Database schema design (Sidereus);

— a graphic interface system to support Database animation (Database Animator);

— a set of tools devoted to user functions design (Application Editor);

— a Project Database.

The architecture of DEEnv is designed in such a way that the generation of the Galileo code corresponding to the state of work may be performed at any time when the designer needs it so that he/she may better understand the Conceptual Model. The absence of particular Conceptual Modeling situations in which the prototyping may only be performed is an important capability of DEEnv. This characteristc enables the designer to feel free to follow his/her thougths flow without careing the rest of the model. In DEEnv this capability which allow to perform anytime rapid prototyping is called *shared prototyping*.

Sidereus

Sidereus (Albano, 1988; Corte, 1989) is an environment which allows the designer to build and validate Database conceptual schemata (see Fig. 2). Sidereus is based on the use of a semantic data model which is homeomorfous to the Galileo

editor enables building conceptual schemata by using a graphic language which has semantic data model features, being the abstract mechanisms proper to the semantic data model directly offered by the tool. At the beginning, the designer defines the classes, their main structural properties, the relationships existing among the classes together with the necessary constraints. Then, he/she may, where needed, adds assertions among the properties of a class, new properties and new abstract types to better specify the class properties.

In any moment, depending on the designer needs, Sidereus generates the Galileo code corresponding to the state of the art. The produced code is divided into two parts; the first part contains the structural declarations of the schema (classes with their properties, relationships, assertions); the second part contains all the operators which handle data in the way described in the model, i.e. the operators always make consistent transitions of the database. It is important to notice that these are the only operators made available to the designer to manipulate the database. In such a way completely consistency is assured between data and actions upon data.

Database Animator

Database Animator is a tool which allows the designer, even the end user, to interact with the database in a simple and clear way. The interaction is performed by means of a graphic interface (see Fig. 3). Database Animator is adaptable depending on the Galileo code (informations) produced by Sidereus. These informations concern the abstract data types belonging the schema and the operators

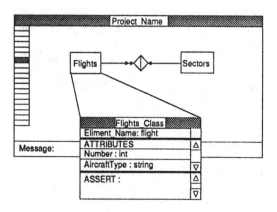

Fig. 2 - The graphic editor

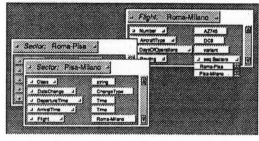

Fig. 3 - Database Animator

language. Sidereus is divided into two components. The first component is a graphical editor, while the second is a Galileo code generator. The graphical

parameters.

The functionalities offered by the tool to the user are

divided into two categories; the first collects the basic operators which perform insertions, deletions and updates; the latter groups the operators to make query not only on the data belonging the database but also on the schema characteristics. Database Animator allows to:

— hang up the current operation;

— begin a new operation after a previous deferring;

— resume an hanged up operation at the same point of deferring;

— finish an operation by an abort or commit action.

Application Editor

Application Editor is a set of tool which supports the designer to build the user functions. It is important to notice that the operators generated by Sidereus are not the user functions but, the bricks which the designer has to use to build them. As a matter of fact, this capability of DEEnv is fundamental so it allows to build only functions which in a consistent way act on the database.

The tools provided by Application Editor are:

— Layout Editor

— Dialog Editor

— Process Editor

Layout Editor. Using Layout Editor the designer defines the interface which each user function has to the end user of the new application. The system asks which are the operators which made up the current being build user function; then it answers with the all parameters needed to allow to work the chosen operators. At this point the designer using a graphical tool builds the interface and its capabilities as the user requires. At the end the Layout Editor generates the Galileo code needed to execute the interface.

Dialog Editor. The Dialog Editor is a tool which supports the designer to build the way by which the end user interacts with each user function. The system allows the designer to associate events and actions to be executed according the semantics of the current being build user function and the state of

the Database.

Process Editor. The Process Editor allows the designer to specify the semantics of each action embedded in the current being build user function.

Project Database

All the informations produced by the designer are permanently stored in a central Project Database. By means of the Project Database the tools of DEEnv exchange data among them. An object-oriented DBMS (OODBMS) has been chosen for the DEEnv Project Database implementation. The need of an OODBMS arises from the following: traditional DBMSs are inadequate to satisfy project databases requirements (Bernstein, 1987); the Project Database gives an *active* support to the Conceptual Modeling. The last assertion requires that exclusive dynamic constraints of the adopted methodology have to be expressed procedurally and this is only possible using an OODBMS as it allows to express the object behavior (Armenise, 1991).

CONCLUSIONS

In this paper is presented an environment (DEEnv) which supports the Conceptual Modeling phase and which achieves the twin goals of completeness and correctness of the design. This is done describeing the design using an executable conceptual modelling language (Galileo) which contains both functional programming and formal specification capabilities. The environment provides graphic tools to support the designers work.

Currently the graphic editor, the database animator and the layout editor are implemented, while the other DEEnv components are being implemented. The implementation is carried out using X-Window rel. 4, and the C language in such a way to reduce the portability problems.

ACKNOWLEDGMENTS

I am grateful to my colleagues P. Corte, M. Diotallevi, D. Presenza, E. Secco and O. Viele for their comments and suggestions on this work.

REFERENCES

Agresti, W. W.(1986). What Are the New Paradigms?. In W. W. Agresti (Ed.) *New Paradigms for Software Development*, IEEE Computer Society Press, 6-10.

Albano, A., L. Cardelli, R. Orsini (1985). Galileo: A Strongly Typed, Interactive Conceptual Language. *ACM Transactions on Databases Systems, 19*, 230-258.

Albano, A., L. Alfiò, S. Coluccini, R. Orsini (1988). An Overview of Sidereus: A Graphical Database Schema Editor for Galileo. In J. W. Schmidt, S. Ceri, M. Missikof (Eds.) *Advances in Database Technology - EDBT '88*, Springer-Verlag, 567-571.

Armenise, P., S. De Panfilis (1991). Conceptual Modelling for Software Development Methodologies. to appear in *Proc. of third International Conference on Software Engineering and Knowledge Engineering*, Skokie, IL.

Auddino, A., Y. Dennebouy, Y. Dupont, E. Fontana, S. Spaccapietra, Z. Tari (1991). SUPER: A Comprehensive Approach to DBMS Visual User Interfaces. rapport de recherche (submitted for publication), Ecole Polytechnique Federale de Lausanne.

Benedusi, P., V. Benvenuto, M. G. Caporaso (1990). Maintenance and Prototyping at the Entity-Relationship level: a

Knowledge-Based Support. *Proc. of the Conference on Software Maintenance*, San Diego, CA, 161-169.

Boehm, B. W.. (1976). Software Engineering. *IEEE Transactions on Computers, c-25*, 1226-1241.

Braegger, D., Z. Rebsamen (1985). Gambit: an Interactive Database Design Tool for Data Structures, Integrity, Constraints and Transactions. *IEEE Transactions on Software Engineering*, 574-582.

Chen, P. P. (1977). The Entity Relationship Approach to Logical Database Design. Q.E.D. Monograph Series, Wellesley, Massachusetts.

Corte, P., D. Presenza (1989). Un editore grafico per la generazione di schemi concettuali Galileo. Tesi di Laurea, Università di Pisa.

DeMarco, T. (1978). *Structured Analysis and System Specification*. Yourdon Press, Englewood Cliffs.

Goodman, A.M. (1988). IMSADF - a tool for programmer productivity. *Database, 11(3)*, 106-113.

Henderson, P.(1986). Functional Programmin, Formal Specification, and Rapid Prototyping. *IEEE Transactions on Software Engineering, SE-12*, 241-250.

SECURITY EVALUATION CRITERIA

S. J. Knapskog

University of Trondheim, Division of Computer Systems and Telematics, Trondheim, Norway

Abstract. Evaluation of security for an information technology (IT) product or -system has become a paramount task to ensure that the end users needs are taken adequate care of. The last couple of years have shown a great improvement, both nationally and internationally, with respect to different initiatives and proposals for standardization of the processes and mechanisms necessary to achieve a proper security evaluation. It should be clear that the ultimate goal in this field must be to reach a world wide accepted agreement on the international level, so that the different national evaluation and certification schemes may be mutually recognized for the full benefit of the free flow of IT products and -systems into the international market. To this end, IEC and ISO have, under their Joint Technical Comittee no. 1 (JTC1), Sub Committee 27 (SC27), established a working group (WG3) named "Security Evaluation Criteria" to work in this field, and to produce an international standard "Criteria for Security Evaluation". Other related areas, such as user requirements, and recommendations to national member bodys on how to organize the evaluation schemes locally, will also be considered in the future standards.

Keywords. Open systems, information technology security, security evaluation, evaluation criteria, certification, accreditation.

INTRODUCTION

The history of criteria for evaluation of the security of information technology (IT)-systems is not a very long one. Evaluation of commercially available computer systems, or more precisely: operating systems, started in the early eighties. The first national standard, the US Trusted Computer Security Evaluation Criteria (TCSEC, or the "Orange Book") was published in 1983. Since then, a lot have happened in the field of IT systems security. Many national evaluation-, certification- and accreditation schemes have been presented, but still there exist no international standards covering this vitally important area. In this paper, the principles of security evaluation will be discussed, and the need for practical schemes for evaluation and certification will be pointed out. The definitions and terms used in this paper are aligned with the Draft Information Technology Security Evaluation Criteria (ITSEC), version 1.1, published by the Four-Nations Group: France - Germany - the Netherlands - the United Kingdom. Towards the end, the status of the on-going standardization work within ISO will be given.

IT SYSTEMS SECURITY

The three basic threats against IT systems security are:

- loss of confidentiality
- loss of integrity
- loss of availability

A secure IT system in this context is a system composed by IT products containing security functionality installed in their operating environment. A secure IT product is an IT system containing security functionality, but the operating conditions are not known (they may be assumed for evaluation purposes).

Security functionality is the result of implementation of security mechanisms integrated in the IT products. It can be further divided into:

a) Security objectives - Why the functionality is wanted

b) Security enforcing functions - What functionality is actually provided

c) Security mechanisms - How the functionality is provided

The following is a list of basic *security functions*:

- identification and authentication
- access control
- accountability
- audit
- object reuse
- accuracy
- reliability of service
- data exchange

The data exchange security functions are defined in accordance with the ISO/IEC Standard ID 7498-2: Security Architecture:

- peer-entity authentication
- access control
- data confidentiality
- data integrity
- data origin authentication
- non-repudiation

Security functionality is realized by implementation of the security mechanisms. These mechanisms are all the technical, operational, procedural and organizational measures taken to achieve the security aims of the product or system. Typical technical mechanisms are encryption and message authentication procedures, whereas the operational and procedural measures encompass physical security of sites and equipment and positive identification of personel before access to a given resource is granted. Organizational security mechanisms should be laid out in a *security policy* for the organization in question.

The level of confidence that can be established that the security functionality of the IT system is adequate, both with respect to *correctness* and *effectiveness* of the security functions will be designated *assurance*. The *assurance level* obtainable will thus also depend on the strength of the security mechanisms used to implement the security functions. The security mechanisms can be given their own rating with respect to strength, like for instance in <ITSEC>: *basic*, *medium* or *high*.

IT SYSTEM SECURITY POLICY

The system security policy specifies the set of rules and practices that the organization will enforce to protect its information and the IT systems that contain and process this information. In some cases, also a nation's laws will influence the necessary precautions taken to protect information. The policy shall identify the security objectives of the system and the threats to the system. These security objectives shall be addressed both by the IT systems security enforcing functions and by physical, procedural and organizational security measures. For a large organization, the IT System Security Policy often will be one part of a multipart Corporate Security Policy, providing the context for the identification of system security objectives for all the different IT security systems. In the other direction in the policy hierarchy, the IT System Security Policy will consist of different parts. In this context it is appropriate to point at the document designated the Technical Security Policy, containing the set of rules and practices for protection of sensitive information stored or under processing in the IT system, and the software and hardware parts of the IT system itself. The IT System Security Policy is the main part of the top level documentation against which the systems security is evaluated.

PREDEFINED CLASSES OF FUNCTIONALITY

Many systems will have similar security objectives, and therefor will be needing the same security functionality, and the security objectives will be met by the same security enforcing functions. A predefined functionality class will meet specific markets sectors common need for security, and give guidance to both manufacturers and users by providing a standard set of security functions. If well known and accepted, i.e. standard implementation methods are used, the level of assurance may be precisely defined.

SPECIFICATION METHODS

The stronger the demand for IT system security, the stronger the required formalism in description, construction and implementation of the IT system. Three levels of sophistication can be identified:

- Informal specification, written in natural language. A specification written in natural language is not subject to any special restrictions, but adheres to normal syntax and grammar rules. Terms used should be well defined, and ambiguities avoided.

- Semiformal specification, requires use of a restricted notation in accordance with some convention that is much more stringent than natural language syntax and grammar rules. One well known semiformal description method is the CCITT standardized SDL.

- Formal specification, written in a rigorous, mathematically based notation. The concept must include strict definition of semantics and syntax, and proof rules supporting logic reasoning. In a hierarchical design cycle, it must be possible to prove consistency between operation at different levels of abstraction. One example of a formal specification language is LOTOS, standardized by ISO.

ASSURANCE - EFFECTIVENESS

The criteria for effectiveness covers several aspects of IT systems security. To be able to evaluate the system, one must perform a threat analysis, and then carefully consider the suitability of the implemented security enforcing functions to counter the identified threats. The strength of the mechanisms used to withstand a direct attack must be taken into consideration, and the mutual supportiveness of the mechanisms and the security enforcing functions. In the product evaluation scenario, the risk of unintentionally compromising the system security by configuring or using the system in a non-authorized way shall be assessed. It is also important to assess whether known security vulnerabilities in the construction or operation of the IT system in fact could compromise the IT system security, and indeed even some of the generic security functions or security mechanisms involved. The vigour of the effectiveness evaluation is dependent on the specified assurance level for the IT system.

ASSURANCE - CORRECTNESS

The evaluation criteria for correctness may contain several levels with increasing demands on detail and formalism of the construction and implementation documentation. The necessary evaluation documentation can be broken down as in the following <ITSEC>:

Development process

- requirements
- architectural design
- detailed design
- implementation

Development environment

- configuration control
- programming languages and compilers
- developers security

Operation

- operational documentation
- delivery and configuration
- startup and operation

The burden for provision of evidence is on the manufacturer or vendor of the IT system, and the evaluators main track is to check the validity of the claims put forward. Testing of security functionality by the manufacturer is one aspect of the quality assurance normally performed in an active Quality Assurance Programme. This Qualaity Assurance Programme itself must be adequate for the desired assurance level for the IT system under evaluation.

A PRACTICAL EVALUATION, CERTIFICATION AND ACCREDITATION SCHEME

The result of an evaluation in accordance with the evaluation criteria as described briefly in this paper, is an established trust that the evaluated IT system really has the claimed security functionality, and may be given an assurance rating to one of the levels defined in the criteria. This fact may be used by the end users in procurement of equipment with adequate security features for performing a specific information processing task, or it may be used by the manufacturer to prove that they have produced equipment containing evaluated security functionality tailored to specific market sectors need. In order to reap the full benefit of an evaluation, it will be necessary to have a trusted, not necessarily public, organization to issue a certificate to the fact that products and systems have been evaluated and achieved a rating in accordance to the standardized criteria. To establish trust in the evaluation process itself, it will be equally necessary to have free standing evaluation facilities licenced by an accreditation process, performed by the certificate issuing, trusted organization. A practical scheme will consist of:

- evaluation sponsors (manufacturers, vendors, end users)
- evaluation facilities (licensed test laboratories)
- accreditation and certification authority

A strict departmentalisation between manufacturers and evaluation facilities is necessary to ensure neutrality, even if a close cooperation during the evaluation will be necessary in most cases. The evaluation facilities on their part will have a great responsibility towards both manufacturers, vendors and end users because an evaluation necessarily will demand access to detailed proprietary and private information about products and production processes.

INTERNATIONAL STANDARDIZATION OF EVALUATION CRITERIA

The ISO/IEC Joint Technical Committee No. 1 "Information Technology" (JTC1) has among its many subcommittees SC 27 entitled " Security Techniques". The subcommittee is divided in three Working Groups:

    WG 1 - Requirements, Security
           Services and Guidelines
    WG 2 - Techniques and Mechanisms
    WG 3 - Security Evaluation Criteria

The Terms of Reference for WG3 are:

1.   Standards for IT security
      evaluation and certification of
      IT systems, components, and

products. This will include
consideration of computer
networks, distributed systems,
associated application services,
etc.

2.   Three aspects may be dis-
      tinguished:
      a) evaluation criteria
      b) methodology for application of
         the criteria
      c) administrative procedures for
         evaluation, certification and
         accreditation schemes.

3.   This work will reflect the need
      of relevant market sectors in
      society, as represented through
      ISO/IEC National Bodies and other
      organisations in liaison, ex-
      pressed in standards for security
      functionality and assurance.

4.   Account will be taken of related
      ISO/IEC standards for quality
      management and testing so as not
      to duplicate these efforts.

The work of SC27/WG3 will be based on
existing national and international
proposals for evaluation criteria and
certification and accreditation
schemes. The future international
standards and schemes for application
of the standards must be based on
National Member Bodies full agreement
so that mutual acceptance of the
different national schemes based on
the international standards is
achieved. This is of paramount
importance both for manufacturers,
vendors and end users of IT systems,
to avoid sectorization of markets and
the serious consequences such as
higher end user costs and less
connectability.

STANDARDIZATION STATUS

The very first draft of parts of the
international standard for evaluation
criteria is due in Oct. 91. The production of international standards is
by nature a slow process, and it can
be foreseen that several WG meetings
will be needed before the proposals for
a standard have reached the necessary
maturity to be unanimously accepted
within the working group itself. There-
after follows a procedural delay,
consisting of balloting rounds to the
National Member Bodies before the
international standard is eventually
published. It is the hope that the
practices established nationally and
by bi- or multilateral agreements in
the meantime can be reasonbly easy
adjusted to adhere to the new standards, as this will speed up the
process of creating a true common
denominator for evaluation and
certification of IT systems security,
both with regard to security
functionality and assurance levels.

CONCLUSION

In this paper we have discussed the
principles involved in evaluation,
certification and accreditation schemes
for security of IT product and systems.
It is evident that international
standardization in this field is
urgently needed to ensure that the end
users security needs may be satisfied
to a well defined level of trust, and
that nationally implemented schemes for
evaluation and certification may be
mutually recognized on a world wide
basis.

REFERENCES

<TCSEC>: Department of Defence Trusted
          Computer Systems Evaluation
          Criteria, DoD 5200.28-STD,
          Department of Defence, USA,
          Dec. 1985.

<ITSEC>: Information Technology
          Security Evaluation Criteria
          (ITSEC). Harmonised Criteria
          of France - Germany - the
          Netherlands - the United
          Kingdom, Version 1.1, Jan
          1991.

INFORMATION SECURITY ISSUES IN TRANSACTION SYSTEMS APPLIED TO AN INTEGRATED ROAD TRAFFIC ENVIRONMENT*

B. G. Andersen

SINTEF DELAB, N-7034 Trondheim, Norway

ABSTRACT: Application domains for electronic transaction systems are increasing in number. Transport is a new area where the information security aspects are of great importance, and where a number of new problems arise due to the mobility aspects.

The paper deals with the pan-european scenario IRTE (Integrated Road Traffic Environment). Within the transport area there is a need for a secure communication between integrated circuit cards (smart cards) in the mobiles and road side charging stations, and between different charging stations within a larger geographical area.

Several issues related to the information security are identified. The paper concentrates upon:

- Secure funds transfer for a pan-european IRTE, covering the payment network

and

- access control/rights to different types of information stored in an integrated circuit card (smart card). This includes several different parties like card issuers, service providers and card owners.

KEYWORDS: Communication computers application, transportation, security, electronic transaction system, smart cards.

1 INTRODUCTION

The objective of this paper is to identify information security problems and aspects within the application of an Integrated Road Traffic Environment (IRTE).

The security aspects are mainly related to the payment service in the IRTE scenario, namely an automatic debiting system including road pricing, parking and public transport like busses and ferries. This new generation of debiting systems will probably be based on integrated circuit cards (smart cards) with memory and possibilities for remote read and write access.

See figure 1 for a functional overview of the debiting system.

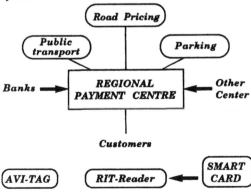

Figure 1. An automatic debiting system.

Within this application the most salient information security problems are:

- identification of card holder (user authentication),
- protection of authorizations in the card (integrity),
- protection of information stored in the card (integrity, confidentiality),
- authentication of the transactions,
- confidentiality and integrity of messages and keys, and
- protection against forged data in connection with communication.

Internationally, specifications of automatic debiting systems have been worked with inside the DRIVE projects PAMELA, TARDIS and SMART. The specifications are quite complex and detailed, but elements like security and anonymity are only dealt with very shortly.

The security aspects are in the paper mainly dealt with from the system's point of view, but anonymity as an important aspect of the users security is discussed as well. The paper starts with discussing the security aspects for electronic transactions systems. The discussion uses as its basis a model for the toll transaction, and the security threats and requirements are outlined in connection with this model. This part encompasses Chapter 2.
The next part of the paper deals with security in integrated circuit cards (smart cards). Different security issues related to the functionality of smart cards and to the user security are outlined. This is covered in Chapters 3 and 4.

---

* The work reported in this paper has been done under contracts with Micro Design A/S.

2 SECURITY ASPECTS OF ELECTRONIC TRANSACTION SYSTEMS

This chapter contains a description of the security aspects for electronic transaction systems. A formal model of the toll transaction is outlined in accordance with the constraints of some of the operational requirements from the PAMELA project in DRIVE. On the basis of this model the security threats and requirements for a toll transaction system are pointed out.

2.1 Operational requirements

The PAMELA report <7> lists several operational requirements where several have implications on the constraints of the payment system design. These are listed below.

- Any driver must, if they want, be able to use the system anonymously.

- The road-use pricing system should be simple for all road-users to understand.

- The system should be reasonably immune to attempts at fraud and evasion, whether deliberat or unintentional.

- The system should be reliable, robust and accurate.

- The system should be pan-european.

- It must be possible periodically to revise toll struture.

- No double charging must take place.

- Different currencies must be accepted.

- Smart cards must not be transferable between vehicles which are charged at different rates.

- The smart card must be designed to protect against fraud either when recharging the card, or when paying for the use of the road.

The requirements show a preference for realtime payment. A key issue is the directness of payment; drivers must perceive the payments.

2.2 A model for a toll transaction

In the sequel we describe a formal model which attempts to abstract all relevant entities and events in toll transactions as well as transactions in connection with public transportation and parking.

Model formulation

The entities involved in the transaction are a roadside charging station ($S \in \mathbf{S}$), a vehicle ($V \in \mathbf{V}$) which is to be charged when passing S, a "coin" ($C \in \mathbf{C}$) which is sent from V to S, and the "receipt" ($R \in \mathbf{R}$) which is sent from S to V. Here, $\mathbf{S}$, $\mathbf{V}$, $\mathbf{C}$, and $\mathbf{R}$ denote the set of all roadside charging stations, the set of all vehicles, the set of all coins, and the set of all receipts, respectively. A "coin" is some piece of information that somehow enables the transfer of real funds from the owner of V to the owner of S. A "receipt" is some piece of information which yields a proof for the owner of V of the fact that S accepted C as valid and permitted V to pass.

We assume from now on, that all entities are unique,

that is, S represents a unique roadside charging station, V represents a unique vehicle, C represents a unique coin, and finally, R represents a unique receipt. The model is presented in form of a function which we call payment (P); P: $\mathbf{S} \times \mathbf{V} \times \mathbf{time} \rightarrow \mathbf{R} \times \mathbf{C}$. A payment is uniquely defined by the S and V involved, and the (unique) time the payment takes place. The outcome of a payment is a coin and a receipt.

Let in the following $\bar{S}$ denote a "straight" S, which follows its prescribed protocols. Let $\widetilde{S}$ denote a "crooked" S, which is faulty and does not necessarily follows its prescribed protocols, some entity masquerading as $\bar{S}$, or some entity eavesdropping the communication between $\bar{S}$ and $\bar{V}$. Let $\bar{V}$ and $\widetilde{V}$ be defined similarly. In addition, let $\bar{C}$ and $\bar{R}$ denote a valid coin and a valid receipt, respectively, and $\widetilde{C}$ and $\widetilde{R}$ denote an invalid coin and an invalid receipt, respectively.

The basic assumptions of the model are that $\bar{V}$ will not pass $\bar{S}$ before it has got hold of $\bar{R}$ after delivering $\bar{C}$.

We claim that our model covers any toll payment transaction system.

Security threats

The following "threats" to the security of such transactions are pointed out:

- $\widetilde{V}$ might be able to replicate $\bar{C}$.

- $\widetilde{V}$ might produce $\widetilde{C}$.

- $\widetilde{S}$ might illegally try to obtain $\bar{C}$ from $\bar{V}$.

- $\widetilde{S}$ might produce $\widetilde{R}$.

- If the payments is supposed to be anonymous, $\widetilde{S}$ might try to find the identity of $\bar{V}$ from $\bar{C}$ which $\bar{V}$ delivered.

Security requirements

The following overall security requirements are necessary to meet:

- $\bar{S}$ always accepts $\bar{C}$ from $\bar{V}$.

- $\bar{S}$ never accepts $\widetilde{C}$.

- $\bar{S}$ never sends $\bar{R}$ to $\widetilde{V}$.

- $\bar{V}$ always accepts $\bar{R}$.

- $\bar{V}$ never accepts $\widetilde{R}$.

- $\bar{V}$ never sends $\bar{C}$ to $\widetilde{S}$ (this includes that no entity must be able to get hold of $\bar{C}$ by eaves dropping the communication between $\bar{S}$ and $\bar{V}$.)

- $\widetilde{V}$ must not be able to neither replicate $\bar{C}$ nor produce $\bar{C}$ on its own. (That is, only the coin issuing organisation must be able to produce the coins, possibly in cooperation with $\bar{V}$.)

- No one must be able to establish the identity of $\bar{V}$ based on information about the triple ($\bar{C}$, $\bar{S}$, time), assuming anonymous payments. (In technical terms; the function f ($\bar{C}$, $\bar{S}$, time) = $\bar{V}$ must be hard to compute.)

Note that even though it is $\bar{S}$ who issues the coins, it must not be able to recognize the coins when they are delivered back by $\bar{V}$.

3 SECURITY IN INTEGRATED CIRCUIT CARDS (SMART CARDS)

In this chapter security aspects related to the use of integrated circuit cards (smart cards) are identified. This use of smart cards may be looked upon as one way of applying the toll transaction model introduced in Chapter 2. Another way of applying the model may be in connection with using a tag based on a radio signal, or in connection with a manually based system.

The chapter deals with security aspects from the system's point of view by mainly focusing on the functionality of the smart card. Security aspects like authentication of messages and cards, access control to read and write operations, and access levels to data storage are discussed. In addition some general security aspects in connection with smart cards are pointed out.

3.1 Security aspects for smart cards

An Integrated Circuit Card (smart card) can be looked upon as a portable device with intelligence and provisions for identity and security. Increased intelligence and processing power becoming available may be utilized by the designer to support the inherent security and identity features. There are several overall security aspects to take into account in connection with smart cards, and some of them may be formulated as follows:

- Absolute security may not be an aim in itself, and in the normal case the designer has to select a level suitable for a given application.

- The card's own security is but part of the total system, which in turn is only as strong as its weakest link.

- System security is a shared responsibility stretching from the supplier, through the system operator/card issuer, to their trading partners or agents down to the individual card holder.

- Technical solution chosen must comply with both human and economic factors and not impose unacceptable or unworkable requirements.

The more security oriented issues in connection with smart cards are:

- **Card authentication:** to confirm that the card is a legitimate card.

- **Terminal authentication:** to confirm that the terminal to be used by the card is not an unauthorized or fake device.

- **Card holder verification** (User authentication): to confirm that the person presenting the card is the authorized card holder.

- **Message authentication:** to verify that the elements of the received message are genuine.

3.2 Functionality of a smart card

In a scenario with multiple functions implemented in the card, and multiple services to be offered from different vendors or organizations, methods for securely establishing and accessing separate areas of the card's memory must be found. Usually, a service provider will insist on the possibility to initialize and personalize the card in accordance with the agreed services paid for. Such actions will involve a card programming device, and a number of unsolved problems exist within this area. Some examples are:

- limitation of write access for authorized persons or entities to defined ("own") storage sectors

- denial of read and write access for authorized persons or entities outside own sector

It will also be necessary to establish the practical arrangements and procedures for issuing, renewing and revoking cards, or given functionality of a multifuntion card.

An Integrated Circuit Card (smart card) is a card with standardized dimensions and placement of electrical contacts, and significant in-card data processing and data storage capabilities. The communication between the card and the card accepting device is regulated in accordance with the ISO report <3>. Both a synchronous and an asynchronous communication mode has been standardized. The protocols used are commonly used in many other applications as well, and there exist standard methods for encryption of data on this level, described in the ISO report <5>.

Basic security of smart cards

There is a general consent that forgery of a smart card is inherently very difficult. Nevertheless, given advanced engineering tools and knowledge, reverse engineering of a card is feasible. Some kind of tamper-proof mechanisms must be used, in such a way that internal structures conveying information of the functionality of the smart card, or vital data such as cryptographic keys, are effectively destroyed if an attack is mounted against the smart card. Especially system owners must be aware of this fact, because even if the value of each card used only for payments of limited amounts are small, the large number of cards in use may make such an attack tempting for a powerful adversary. There are a number of technical problems involved with tamper-proof designs in a limited volume such as the standardized credit card format <1, 2>.

The availability function is dependent on a card functioning correctly at all times. Growing functionality and adding of security functions demand physically larger chips to be put into the card. There is a distinct possibility that this will constitute a violation to the standardized requirements for physical flexibility and durability of the smart card. There are two ways of combating these problems. Either, the physical restraints are to be met and the limitations on functionality and security of the card are accepted, or the physical dimensions of the card must be allowed to deviate from existing standards with resepct to size and shape.

Authentication of messages

Given that a future smart card is a self-contained unit with display and keyboard, standardized methods for authentication and encryption of messages exist, and they can readily be implemented in the card <4>.

All the well-known, partly standardized, features of a
secure communication link can be exploited, including
a hierarchical scheme for distribution of cryptographic
keys, and a security management scheme for admini-
stration of the security functions of the total system.
Algorithms used, both symmetric and asymmetric,
should preferably be publicly known and registered, so
that the strength of this part of the system can be
evaluated <3, 4>.

Security of card readers (card accepting device)

In the scenario outlined above, the card reader is only
relevant to communication; it contains no specific
security functionality.

However, if the card is not completely self-contained,
because it does not have a display and a keyboard, the
picture changes somewhat. In this case, the terminals
(card readers) must contain some security relevant
information, and must be secured accordingly, with a
tamper-free design of vital parts. In such a scenario,
the terminal must be considered the weak part of the
link, and emphasis must be put on its security features
accordingly.

Authentication of card to terminal

Assuming a card, the most usual case is for the terminal to be supplied with an algorithm which matches
that already resident in the card. The terminal issues a
"challenge" to the card, which computes a result using
its matching algorithm and returns this response <6>.
The terminal microprocessor conducts a calculation
and compares its result with that returned by the card.
If the two match, this should ensure that the card
presented is genuine.

The terminal operator has to be certain that the
authorized terminal has not been exchanged for an
illegal version designed to intercept and re-direct
valuable inputs. The secret data necessarily resident in
order to conduct "challenge and response" procedures
during the operational period when each card is presented, must be removed and protected whenever the
terminal is disabled.

The terminal in itself can be viewed as an end-node in
the network, and it has similarly to be authenticated to
the network.

Access types and access levels

As the technology progresses, the amount of data
capable of being stored in the card will increase. Due
to this there will be an extended need for access control
to the system and to the information.

In addition to this several actors will be involved. The
actors can be divided into three main groups:

1)   Card holders

2)   Service institutions (toll road companies, car
     parking companies, public transport operators,
     banks, financial institutions, etc.)

3)   Control courts (police, vehicle authority, etc.)

The two main access types are READ and WRITE
access. For each of these types there may be different
levels of access to the information, depending on what
kind of operation that is to be carried out.

In connection with READ access it may be necessary to
make a distinction between detailed information and
more general statistics. The actors in these applications
may need access to different kinds of information, and it
should be controlled in such a way that each specific
actor has access only to the set of information specified
for this actor. This includes both type and level of information.

In connection with WRITE access the distinction may
be done between different write operations, like add/
insert, overwrite/modify and delete.
The different actors should have different kinds of write
access, depending on the type of write operation.

Data storage

To be able to handle the problems connected to multi-
actors and data access, the data storage has to be
divided into different logical areas, each with different
types and levels of access. The solutions to this may be
found by using concepts and methods taken from access
control lists and capabilities. There will also be a need
for cryptograhpic keys; the secret key located in the
card and the public key in the terminal.

In this paper we have focused on the data storage in the
smart card. The information stored in the card may also
be stored at other sites in the system, and the information stored in the card may as well be used in combination with information stored at the other sites in the
system. This creates further problems.

4 USER SECURITY

The future, self-contained smart card will generally
contain a lot of information that the average user will
regard as sensitive, both of personal nature and more
technically sensitive, as for example cryptographic keys.
The user need to be convinced of the fact that the
security measures taken will be of sufficient strength to
protect this information. Once that is established, the
user himself will be in full control of the card and of the
information contained in the card.
In the intermediate phase, one will have to accept that
some security functionality is implemented in the card
readers (terminals), and those terminals will be out of
the users immediate control.
It is therefore necessary to establish a condition of trust
between the users and the organizations responsible for
the construction and maintenance of the terminals.
Given that some such organizations have already established some level of trust in relations to the general
public (i.e. banks, Electronic Funds Transfer (EFT)
systems), this should not be very difficult. The general
acceptance of use of plastic cards for different kinds of
transactions, and of the technical equipment embedded
in existing transaction systems is high. This attitude
must be taken care of. The main remedy to obtain this
is to be able to define a clear and comprehensive security policy both for the security domains for the separate sub-systems, if such exist, and for the total system
as seen by the user.

4.1 Multifunctionality

Some points have already been made of the fact that the demand for multifunctionality of the cards increases the risk of unauthorized manipulation of the card's memory. For the users, this is an obvious threat. When the card is delivered to the owner of a sub-system for "recharging", there is little or no possibility for the user to control or monitor the process, and thus be convinced that the sub-system owner only performs authorized changes of the card's functionality or security. Some kind of agreement or contract must be put in place to regulate this process and mutually define the responsibilities of each participating party.

4.2 Authentication of user to card (card holder verification)

Todays user authentication is solely based on PIN-verification. This is a very convenient method, but gives limited security. In the future the authentication probably will be based on either biometrically funded methods, or even more likely, a combination between physical characteristics of the user and a PIN-number or similar code. It is generally accepted that a self-contained card that can perform the whole authentication sequence locally, is to be preferred from a security point of view. Existing solutions based on storing cryptographic keys for decryption of the PIN code in the terminal enhances the vulnerability of the system immensely.

4.3 Anonymity of users

The obvious solution to this question is to implement systems with pre-payment of services rendered. There are, however, several arguments against such solutions in a general multifunction scenario. To obtain anonymity in an on-line or a post-payment system, crypto-graphic protocols based on public key cryptography can be used, albeit at the expense of considerably higher traffic volume and technical and procedural complexity in the terminal or smart card itself. There is also the question of general acceptance of the principle of anonymous cryptographic protocols, exploiting the concept of "electronic cash" to its full potential. This may, however, be added to the system at a later stage when the understanding of the principles and the technology have become more mature.

5 FURTHER ISSUES

In the following, we list some general security issues to be further investigated and pursued into specific implementation proposals.

5.1 Electronic transaction systems

Cryptographic protocols

When the constraints of the toll transaction model have been determined, the specific cryptographic protocols must be constructed.

Constraints and consequences

The technical constraints must be determined, and its consequences with respect to:

- real time constraints
- storage capacity in driver token and backbone system
- channel capacities and errors
- cost
- reliability

must be identified.

Inter-station cooperation

In order for cooperation among several toll stations to take place, the specific interfaces and protocols that need to be agreed upon must be identified.

Trade-off

The advantages and disadvantages of:

1) pre-payment
2) digital cash
3) post-payment

with respect to

- technical constraints (as above)
- cost of security mechanisms
- clearing and currency problems
- operational requirements set forth by PAMELA and national legislation
- user acceptance
- social and political aspects

must be clarified.

5.2 Integrated Circuit Card security

User authentication

As indicated in section 4.2, there are clear indications of the fact that new methods should be found for this function to enhance the security achieved with todays PIN-based routines. Encryption protocols for the PIN (or equivalent identity information) must be established as close to the user as practical.

Terminal authentication

As mentioned in section 4.3, the terminal, personalized through the use of the smart card, must be authenticated to the network/service providers. This can in principle be done according to existing "standards", but the realizations are critical with respect to data transmission rate and memory usage, dependent on which algorithms that are chosen.

Memory compartmentalization and programming of smart cards

Each smart card will contain data and procedures specific to each service provider. The initialization and recharging of the memory part allocated to each service provider will normally be performed at the service providers terms (and perhaps even premises). This poses a great challenge to the internal security mechanisms of the card.

System security

Efficient implementation of system security functions such as

- generation and distribution of cryptographic keys
- encryption
- integrity enforcement

- audit recording
- violations handling/recovery
- physical security (tamper free-ness)

will have to be studied, modelled and constructed.

6 CONCLUSION

In the balance of this paper we have identified several problems, outlined some general approaches & methodologies, and presented a few specific solutions. Much work still remains to be done, and it has to be done - as the importance of security in such mobile information systems increases continuously.

REFERENCES

<1> International Standard ISO 7816-1. Identification cards - Integrated circuit cards with contacts-Part 1: Physical characteristics. First edition 1987-07-01.

<2> International Standard ISO 7816-2. Identification cards - Integrated circuit cards with contacts-Part 2: Dimension and location for the contacts. First edition 1988-05-15.

<3> International Standard ISO 7816-3. Identification cards -Integrated circuit cards with contacts-Part 3: Electronic signals and transmission protocols. First edition 1989-09-15.

<4> International Standard ISO 9992-1. Financial transaction cards - Messages between the integrated circuit card and the card accepting device-Part 1: Concepts and structures. First edition 1990-06-01.

<5> International Standard ISO IS 9160. Physical layer Interoperability Requirements.

<6> International Standard ISO/DIS 10202-1. Financial transaction cards - Security architectures of financial transaction systems using integrated circuit cards - Part 1: Card life cycle. 1989.

<7> PAMELA ME 1. Strategy for integrated demand management. The case for roaduse pricing.

EXPERIENCES IN DESIGN AND DEVELOPMENT OF A HIGHLY DEPENDABLE AND SCALEABLE DISTRIBUTED SECURITY, ALARM AND CONTROL SYSTEM

E. Schoitsch, W. Kuhn, W. Herzner and M. Thuswald

Austrian Research Centre Seibersdorf (ARCS), A-2444 Seibersdorf, Austria

ABSTRACT: The PHILIPS CSS-System (Compilable Security, Alarm and Control System) is developed and maintained by the Department of Information Technology of the Austrian Research Centre Seibersdorf, which is active in the field of designing and developing software for systems with high safety, reliability or availability requirements, as well as in the process of IT-standardization.

Many dependability issues had to be tackled with in designing the system. The primary goals of the distributed security system are high availability, reliability and maintainability (especially scalability, ie. open to many types of system architectures within any topology (network, redundancy, diversity), and functional extendability). Maintenance problems arise especially with modifications in a distributed and redundant/diverse environment.

Additionally, safety and security issues have to be considered carefully, depending on the type of application. The system combines multiprocessing with functional distribution, configurability with a high degree of software ergonomy ("safe" human interface, reducing human mistakes in operating, maintaining and configuring an application by the customer/user). Standards have to be applied on all system levels where such are available – and applicable in practice !

Keywords: Security, redundancy, dependability, software diversity, system architecture, maintainability, scalability, distributed systems, (software) standards, software development.

INTRODUCTION

The distributed security, alarm and control system called PHILIPS CSS-System (sCaleable Security, Alarm and Control System), which is under development for Austrian PHILIPS Industry AG, has as primary goal high availability and scalability (ie. configurable freely within any topology (network, redundancy and if necessary diversity), security, credibility and integrity, and combines multiprocessing with functional distribution, configurability, a high degree of software ergonomy and the application of standards on all system levels where such are available.

In general terms, the primary goal of a security, alarm and control system in general terms is to protect an area, a plant or a building complex, its interior and the people within its boundaries from (any ?) thread from the environment or - hopefully - from inside too - and/or vice versa (depending on the application: Is the outside world more dangerous to the "system" (bank) or the sytem more dangerous to the ouside world (chemical plant, nuclear power plant), and from where is the thread most likely to come ?).

The properties of such a security, alarm and control system are depending on its ability to get information about the environment and its inner status - i.e. the peripheral (sub)systems, sensors etc. - and on the trustworthiness of the (central or distributed) control computer system.

The problem of the peripheral (sub) systems will not be discussed in detail here, the main topic of interest is the control computer system, its system architecture and its software architecture.

For hardware-related systems, there has been for a long time awareness of the problem of quality assurance, and trusted proofs of safety, reliability, availability and related properties do exist for a long time. Software was not trusted as much in critical applications, and fall-back systems which could be trusted had to exist and parallel.

It took some time, until good practices in software design, engineering and methodology emerged and became standardized (ISO 9000-series, IEEE standards, ESA Software Engineering Standards, EWICS TC 7 guidelines, IEC (Software-related) Standards for Safety Related Systems, ITSEC etc.).

Nevertheless, some aspects such as the qualification of systems using redundancy and/or software diversity, systems distributed functionally or geographically, network dependability (especially concerning security) are still not resolved.

One of the problems for introducing new ideas typical for software-controlled systems have been older, hardware-oriented standards using the same terminology which cannot be applied one-to-one to software. On the other hand, some of the new views on software and systems including the systems architecture, will have some feedback on software quality assurance models and software lifecycle models (ESA, 1987).

STANDARDIZATION

Standardization ist one of the tools to improve the quality of a product. Quality attributes affected are availability, portability, maintainability, scalability, acceptability, adaptability, integrity, security, credibility, safety, reliability etc. Most of these system properties are now summarized under the umbrella term "dependability" (IEC/TC 65A/WG 8 document on "System Dependability").

But even nowadays, with software being the basis of information technology (IT), the importance and impact of standardization seems to be at least a little bit questionable in practice. There is an urgent need to define generally accepted terms and definitions, guidelines and procedures as well as strictly standardized syntax and semantics of programming languages, tools and system software interfaces - and a need to be able to enforce the use of these standards to gain full advantage of their use: to increase productivity and quality (EC-TCCB, 1987).

To create such standards which fulfill the criteria mentioned above, guidelines and standards have to be tested in practice on feasibility and applicability before they become obligatory - and then standards may be used to improve the overall system dependability, which means improving the relevant properties of the system. Dependability is the synthesis of these properties. The synthesis model depends on the architecture of the system (Schoitsch, 1989b).

Standards which can be used in practice should

- have a reasonable lifetime
- based on state-of-the art project work
- be tested in practice on applicability and feasibility (field trials)
- not be only a "secondary" of older work even if technology has changed
- be reasonably adaptable to special needs, levels of severity or system architectures if necessary
- not be triggered mainly by commercial interests.

People of our research centre have encountered some of the problems mentioned above in several fields in which they are active in standardization work (software, graphics, telecommunications etc.), and some of those problems have been encountered during the project described here too.

```
            +--- reliability
            |
            +--- availability
            |
dependability--+--- maintainability
            |
            +--- safety
            |
            +--- security
```

Fig. 1: Dependability and (some) of its constituent properties

THE DISTRIBUTED SECURITY, ALARM AND CONTROL SYSTEM

The basic functional levels are shown in Fig. 2, any combination of elements in any configuration should be possible to generate ("scalability"), i.e. starting from a single workstation including all functionality up to a n-fold redundant network with computers of different capability and with functional, geographical and load distribution of high availability. The high degree of flexibility, necessary to meet highly diversified customer needs implies a very high degree of maintainability, which is the ability to maintain the service of the system although modifications have to be integrated or new subsystems added.

Figure 2 visualizes the (logical) architecture of the PHILIPS-CSS system.

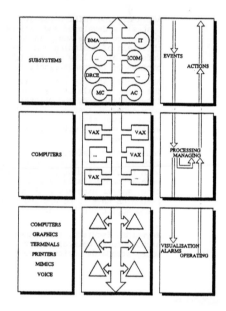

Fig. 2: Open, scalable distributed security, alarm and control system: logical architecture.

All features which may cause the need to choose a distributed systems approach have to be fulfilled:

- safety (reliability, redundancy and/or diversity)
- functional distribution
- geographic distribution
- processing capacity and response time restrictions
- functional autonomy, security

Therefore a thorough study had to be accomplished at the beginning defining goals and non-goals and lining out possible solutions. Goals were:

- flexibility
- open operator (human) interface
- ergonomy of operators workplace
- (human) language independency
- open subsystem interface (functional)
- redundancy (n-fold)
- single-master multi-hot stand-by architecture on a network
- stability (immediate take over)
- maintainability
- security ("secure ethernet" if considered necessary, access control by software- and hardware keys, encryption)
- standards

We are obliged to implement any standard that is available on the hardware and operating system (VAX/VMS). This includes up to now Extended PASCAL (provision to carefully select extensions), SQL, OSI-network, Ethernet (multiple if possible on VAX), X-Windows, FIMS Standard (forms management system), GKS/PHIGS graphics, POSIX-style of using the operating system VAX/VMS, IEC Standards, EWICS Guidelines, ESA Software Engineering Standard - but not all of these standards could be effectively implemented because of reasons like performance drawbacks or implementation problems.

Tasks of the CSS-System

The essential tasks of the system are:

a) to give the operator the view of the outer world he/she needs and wants
b) to give the operator a tool to control different systems with the same method
c) to provide communication-channels between different categories of subsystems of different vendors
d) to provide a long-term storage of incoming and outgoing events for later evaluation
e) to guarantee an extraordinary high availability if needed (dependent on the system architecture choosen)

System Design

The system was designed using PROMOD as a CASE tool. Though PROMOD provides the functionality down to code generation, only high level design (up to 4-5 levels) was used, because of limitations in the PROMOD software. A major goal for the system was the dynamic instantiation of (parallel, identical) processes (e.g. a variable number of operator stations/human interfaces), but PROMOD supports only a fixed number of functional units. In spite of these limitations, it was very useful to use the high level design phase to get an overview about the basic data flow in the system and the consistency of the design.

System Structure

The basic elements of the system are PROCESSES. Each process has a dedicated task, and each process can communicate with any other process. The processes are divided into two classes.

1.) "central" processes
2.) "peripheral" processes

"Central" processes are always present; "peripheral" processes are optional and not limited in number. A typical "central" process is the database administrator. "Peripherals" are operator stations (human interfaces HI, operated in manager mode or operator mode), subsystem drivers, or graphic programs.

All "central" processes must reside on one CPU, but each "peripheral" process can run on a separate CPU. Of course, in "hot-stand-by" respective "Master/shadow" operation modes the set of "central processes" resides on each potential master. The effect of this software architecture is, that the system is most flexible in the number of operating stations and subsystems and that means flexibility with respect to the distribution of resources or extension to new subsystems . No software modification has to be done to add a new operator station.

Fault Tolerance is achieved by active processes, each with redundant "shadow-processes". Additionally, design and software diversity on all levels (Schoitsch, 1989b, 1990) may be introduced if an even higher degree of dependability has to be achieved, especially with respect to safety.

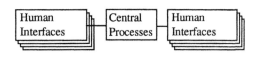

Fig. 3: System Structure: Human Interface

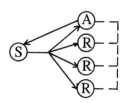

Fig. 4: Fault tolerance and redundant processes (master/shadow principle)

Process "S" sends requests (and information) to all relevant processes "A", "R". All processes R (redundant) have now the same state as "A" (active).

Information exchange

One process in the system can communicate with any other process in the system. Information is exchanged via messages, the system does not use any "shared regions" because of network distribution. A message handling tool has been implemented, which uses either mail boxes (or similar elements available on the respective system) or network services. The communication-path in the operating system is transparent to the processes. It is possible to change the "message layer" or to extend the "communication layer" without changing the "application layer".

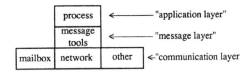

Fig. 5: Software layers (message exchange)

The message send/receive mechanism (input, receive) via mail box is event-triggered. Each process has one and only one input-channel, from where it gets all its information. (FIFO-queue). The message tool provides (up to now) only a one-to-one mapping (one sender, one receiver), multi-cast and broadcast mechanisms are under development.

SUBSYSTEMS

The primary sources of information in the system are the subsystems. Subsystems are typically autonomous systems (hardware, other computers, often other vendors, other software), which can communicate (RS232, Ethernet e.g.) with a host computer. In the security system, each subsystem has a corresponding subsystem driver (a process), which

a) performs communication with the outer world
b) translates the subsystem-specific information into a generalized form (that is the same for all types of subsystems) and v.v.
c) supervises the external links (e.g. line faults)

This model quarantees, that the implementation of a new subsystem (-driver) is a relatively simple task. No changes in the other parts of the system have to be done. Up to now 15 subsystems have been implemented.

- COMMEND Intercom System
- HAUSHAHN Elevator Control System
- INTERFLEX Access Control System 5060+
- PHILIPS Intrusion Detection LAC/TLC
- PHILIPS Video System DRC-E
- PHILIPS Intercom System M100
- PHILIPS Audio System SM40-Routing
- PHILIPS Audio Surveillance System SM40
- PHILIPS Programmable Controller PC20 and MC30
- PHILIPS AMC Keyreader
- SIEMENS Intrusion Detection ITxx
- SIEMENS Fire Detection BMSxx
- SIEMENS Fire Detection SRS450
- TELENORMA Intrusion Detection
- ZETTLER Fire Detection System

In addition, security systems have special functions, which are not performed by just one physical subsystem. Such functionalities are e.g. "security guard-round control", "Door control algorithms", "Camera Connect Algorithms". These "algorithmical" subsystems have the same effect as any other subsystem, but need a combined input/output information from several independent hardware subsystems. The effect of the system architecture is, again, that the system can easily be extended to make use of such algorithms, which combine several independent hardware systems to one new "logical subsystem". (This is as easy as adding new subsystems).

REDUNDANCY

The system may be "compiled" for just one single workstation in small applications up to a multi-computer distributed system implementing single host - multi hot standby redundancy or diversity.

More than one computer may be used for several reasons:

- fault tolerance
- client/server concept
- several workstations for operators, managers, remote site, load sharing etc.
- data back-up/shadowing

All of these concepts are implemented or under implementation, and may be combined. Transition times for master/shadow takes only some seconds, no messages are lost. The client/server configuration allows fast response times for the human interface and fast display build-up. In case of support of several workstations in a configuration, a take-over of "central" functionality by workstations (may be with reduced performance) is possible.

A functional take over may be scheduled at certain times too (e.g. a remote subsystem is operated by an operator during normal working hours, and remote by a primary system overnight). In the most distributed configuration, each subsystem and each operator station may be driven by just one computer each in the network.

MAINTAINABILITY

Maintenance is the preservation or improvement of a systems ability to deliver the specified service during the operational life time of the system. This does include not only fault removal during the operational life of a system, but also adaption to changing needs (enhancement), functionally or topological ("scalability", "portability") (Laprie, 1986).

Maintainability (as a measure) of a system is dependant upon the maintainability of individual parts and the physical, logical and functional structure (architecture) of the system (IEC TC 56, WG 8 draft 1990). This affects ease of access, replacement, ease of diagnosis and ease of adaption/modification.

A basic checklist on maintainability (fault remova aspect) has to include the following provisions to be considered during system design:

- realization of a fault (alert message, report)
- access (modularity, identification, standardization)
- diagnostics, remote maintenance, fault identification
- repairability/replacability
- (self) checking, guided maintenance procedures

Note: most of those aspects have a positive or negative correlation to security aspects of the system too !

System maintenance includes all features quaranteeing a long lifecycle, especially with respect to the most expensive part of the design, the software:

- adaptability
- portability (standards, modularity)
- scalability
- functional extendability

These properties are necessary to be able to change the underlying hardware - starting from adding new subsystems (functional interface !), upgrading the computer system, changing the (network) topology and geographical or functional distribution and ending up with changing the computer type and operating system !

An even more important issue is maintainability in a distributed system, where this property has to be designed into the system from the beginning - it cannot be introduced later. As basis we use the EWICS/IEC guideline (EWICS TC7, 1989), adapted (as "field test" as described earlier) to be applicable to the problems of a redundant network. Problems arise especially with modifications:

- how to modify consistent functionality in a scalable network ?
- how to modify consistent functionality in a redundant system ?
- how to modify/extend consistent functionality in a diverse manner ?

These problems had to be considered. The rules and procedures of the existing EWICS guideline on maintenance and modification have to be carefully checked on their applicability in case of flexible architectures. Especially the system properties "redundancy", "distributed and scalable" and (later on) "diversity" shall not be affected in their effect. It should not be forgotten, that such an emergency system has some safety related aspects too (eg. closed firedoors and people inside the fire zone). Adaptions/enhancement must be done in the same manner as the original architectural design was done - else even small changes may insert "single point of failures" into the system and degrade the dependability of the whole system. This has to be checked via checklists based on the original design.

The results of our work on this project surely will influence our contribution to (pre-) standardization bodies like EWICS TC 7 who are now preparing guidelines on topics such as "Safety Aspects of Distributed Systems" or on security aspects respectively maintenance. On the other hand, the exchange of expertise on this topics is very valuable in learning and to check our concepts on feasibility. The result could never be the same on a mere theoretically level.

SECURITY ASPECTS

VAX/VMS as an operating system has been chosen because it provides more features and user-comfort as well as possibilities of system shadowing and dual access options for devices and the cluster features, which allows a view of a distributed computer system as just one system with all resources shared (CPU, storage) - and because it is (at the moment) more secure than UNIX systems.

VAX/VMS has been evaluated by the National Computer Security Centre, USA, (NCSC, 1986) and has been classified for C2 (overall evaluation class) fulfilling B2-trusted path requirements additionally. This includes priviledge and protection requirements. If required, a B1-version SEVMS version would be available if the US-license is granted for that country and application.

The following security mechanisms are supported in the CSS-application:

- User access with

  a) User name/password
  b) User name/ 2 different passwords (two persons access)
  c) Card, key, etc.
  d) Two cards, two keys, etc.
  e) Combination of card/password

- Priviledge based access for each functional unit of the system
- General security mechanisms of the underlying operating system VMS (user access, file- and database multi-level protection schemes).
- Optional encryption of data which are sent via a network-path.
- Logging and storage of all user activities (including data base changes, hardware & software error logging, security breach detection, network activities)

HUMAN INTERFACE

The ergonomic aspects of the human interface have to be designed in a "safe" manner (reducing human mistakes in operating, showing inconsistencies of the system too). To accomplish this goal, extensive prototyping in the early design phase combined with provisional "dummy" use (only output of the actual system operational) has been done. Ergonomic aspects have been studied too - to provide appropriate icons, pictograms etc. to improve user friendlyness and safety of use (Staufer 1987), which of course can be changed or created by the user too.

Standards have been used on all system levels where available and possible from the performance point of view. Therefore FIMS (CODASYL Forms Interface Management System) in the Digital implementation called DECforms was used as

starting point. Unfortunately, we had troubles with simultaneous, asynchronous input/output as necessary for alarm messages which is not supported in the standard. Therefore this task was separated into two tasks:

- development of a form description language based on a subset of FIMS
- separate definition of layout/functionality for display and the interaction with the operator such implementing asynchronous I/O

Further requirements were:

- configurability/adaptibility to customer needs
- independent of natural language (text configurable)
- independence of hardware, windowing system (X-windows now, change to OSF/Motif possible)
- flexibility/ease of use
- forms description application and system independent
- device-independence through device-dependent modules (interfaces)
- use of thumb terminals, of keyboard or mouse possible

All objects may be handled by novice use as well as by expert users, supplying different modes of operation (menues and short-cut commands). These requirements were fulfilled mainly by sticking to standards and choosing an appropriate software architecture.

The color graphics system of CSS, as all other modules, is completely written in Pascal, where the international standard IS7942 "GKS - Graphical Kernel System" is used to handle the application-independent respectively device-dependent parts.

GKS not only provides a hardware-independent interface for the programming of graphics applications and therefore allows a significant degree of flexibility in the choice of used graphics-hardware (although a complete hardware-independence cannot be achieved due to basic aspects like the availability of colors or input devices), it also takes over special tasks for graphics data processing, like transformations, rendering, picture storage and structuring, or control of input devices. The used GKS-implementation is that of Digital Equipment.

The graphics 'sub'system of CSS is realized as one process. This process contains both the editor and the alarm display part as independent modules, sharing common subtasks. This supports a fast switch from the editor to the alarm display mode.

Per node/CPU in the CSS-system (distribution via a network!), one such graphics process may be activated, which results in up to one editor and alarm-graphics display per node/CPU.

The use of X-windows instead of a hardware-vendor specific graphic subsystem (Digital VWS/UIS) brought many performance problems. Even an increase in CPU-power by a factor of four, which originally was considered to be sufficient to cope with the overhead of X-windows plus GKS on the workstations used, did not be sufficient. Therefore, a user interface using only standard software tools will be used only with even faster workstations to guarantee the performance demande by the customer.

LIFE CYCLE MODEL

The ESA Software Engineering Standard (ESA, 1987) has been selected by our department for such complex applications like the Security and Emergency System, because it shows a significant advantage over most other software enginering standards: It takes into account the system architecture as important design phase by integrating into its lifecycle such a phase (with all consequences: activities, documents and management procedures). This lifecycle model is supported by tools developed at our department (document lifecycle, analysis tools, regression test tools) and by tools supplied by the hardware supplier Digital (compilers, debuggers, network facilities) or third party vendors such as the GEI (ProMod for structured analysys/design).

The ESA Software Engineering standard is a positive example for a standard. It is relevant for software quality issues as well as for highly dependable critical applications (systems). Older software life cycle models such as the original model of Boehm take into account only software aspects, not system-aspects, and the development- and testphase contains most of the life cycle stages, whereas newer models, applied in fields where safety and reliability are relevant issues, emphasize the system aspect in early stages of the life cycle. The ESA model defines a user requirements definition phase, a software requirements definition phase and an architectural design phase, whereas all stages of development, tests etc. are contained in just one phase – a very realistic model for large, critical real-time projects !

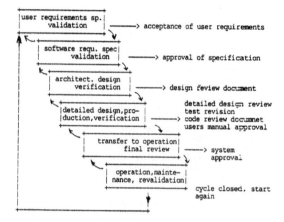

Fig. 6: Variation of the waterfall model according to ESA

CONCLUSION

A project which implies a high degree of availability, security, reliability, scalability and maintainability, has been presented.

The system structure and the architectural aspects as key issue, and their impact on the relevant dependability properties have been demonstrated, together with some implementation issues.

The system combines multiprocessing with functional distribution, configurability with a high degree of software ergonomy ("safe" human interface, reducing human mistakes in operating, maintaining and configuring an application by the customer/user). Standards have to be applied on all system levels where such are available – and applicable in practice !

The problems encountered during the design, development and installation process for such a large, complex distributed system design as well as with maintenance have been discussed and some solutions presented, both in connex with the dependability issues. Some experience with the applicability of standards/guidelines (ESA, IEC, EWICS, ISO) and tools available respectively

added and the problems encountered have been presented too.

Maintenance problems arise especially with modifications in a distributed and redundant/diverse environment.

The system has already been installed at three customer sites by PHILIPS (peripherals and hardware) and ARCS (Software) (one very small and two very large configurations), several others are already ordered.

REFERENCES

Alagic, S., (1986), Relational Data Base Technology. Texts and Monographs in Computer Science, Springer-Verlag, New York.

Avizienis, A., Laprie, J.-C., (1986). "Dependable Computing: From Concepts to Design Diversity" Proc. of the IEEE, Vol. 74, Nr. 5, May 1986, p. 629-638

P.G. Bishop (Ed.), (1990). Dependability of Critical Computer Systems, Vol. 3, Techniques Directory, Elsevier Applied Science, London, New York.

DIN, (1986) Graphische Systeme der Informationsverarbeitung: Graphisches Kernsystem (GKS) – Funktionale Beschreibung. Dokument DIN 66 252, Teil 1.

DIN, (1988) Bildschirmarbeitsplätze – Grundsätze ergonomischer Dialoggestaltung. Dokument DIN 66 234, Teil 8.

EC-TCCB (1987).Forward with Harmonized Conformance Testing Services (EC/TCCB). CTS-Booklet, TCCB Technical Secretariat, Copenhagen, 1987.

Encarnacao, J.L., Encarnacao, L.M., Bono, P., Herzner, W., (1990). PC Graphics with GKS. Prentice Hall, Herfortshire.

ESA, (1987). Software Engineering Standards, January 1987

EWICS TC7, (1989). Reliability, Safety and Security, "Guidelines for the Maintenance and Modification of Safety Related Computer Systems", Position Paper 7.

Frantz, D., Povey, T., (1988). The Forms Interface Management System (FIMS), A Proposed Industry Standard. DECUS Europe Symposium Cannes, 1988.

Görz, G., (1988). Strukturanalyse natürlicher Sprache. Addison-Wesley Verlag, BRD, Bonn.

Hatley, D.J., Pirbhai, I.A. (1987). Strategies for Real-Time System Specification. Dorset House Publishing, New York.

IEC, SC65A, WG9 (1990). "Software for Computers in the Application of Industrial Safety-Related Computer Systems", Standard Draft, 1990.

IEC, SC65A, WG9 (1989). "Functional Safety of Programmable Electronic Systems: Generic Aspects". Working paper 44, 1989.

ISO, (1985). Information Processing Systems – Computer Graphics – Graphical Kernel System (GKS), Functional Description. Document ISO 7942.

Koblitz, N., (1988). A Course in Number Theory and Cryptography. Graduate Texts in Mathematics; 114. Springer Verlag, New York.

Laprie, J.-C., (1986) "The Dependability Approach to Critical Systems". Proc. SAFECOMP '86, October 1986, Sarlat, France.

National Computer Security Centre, Report CSC-EPL-SUM-86/004. Evaluation of VAX/VMS, Maryland, USA, 1986.

F.J. Redmill (Ed.), (1988). Dependability of Critical Computer Systems, Vol. 1, Elsevier Applied Science, London, New York.

F.J. Redmill (Ed.), (1989). Dependability of Critical Computer Systems, Vol. 2, Elsevier Applied Science, London, New York.

Schoitsch, E., (1988). Software Safety and Software Quality Assurance in Real-Time Applications.
Part 1: Software Quality Assurance and Software Safety (Concepts and Standardization Efforts). CPC 50, (1988), 169-188
Part 2: Real-Time Structures and Languages. Computer Physics Communications, 50 (1988), 189-211 North Holland, 1988.

Schoitsch, E. (1989a). The Impact of Standardization on Software Quality, Systems Safety and Reliability, in: Hardware and Software for Real-Time Process Control, Proceedings of the IFAC/IFIP Conference Warsaw, p. 175-186, North-Holland 1989.

Schoitsch, E. (1989b). The Interaction between Practical Experience, Standardization and the Application of Standards, in: Safety of Computer Control Systems 1989 (SAFECOMP '89), Proceedings of the IFAC/IFIP Workshop, Vienna, Austria, 17-24, Pergamon Press Oxford, 1989.

Schoitsch, E. et al. (1990). The ELEKTRA Testbed: Architecture of a Real-Time Test Environment for High Safety and Reliability Requirements , in: Safety of Computer Control Systems 1990 (SAFECOMP '90), Proceedings of the IFAC/IFIP/SARS Symposium, Gatwick, UK, 59-65, Pergamon Press, Oxford, 1990.

Staufer, M.J., (1987). Piktogramme für Computer: Kognitive Verarbeitung, Methoden zur Produktion und Evaluation. Mensch – Computer Kommunikation 2. De Gruyter Berlin, New York.

Theuretzbacher, N. (1986b). Using AI-Methods to improve Software Safety. IFAC SAFECOMP'86, Sarlat, France.

Wirthumer, G. (1989). VOTRICS – Fault Tolerance Realized in Software, SAFECOMP '89, Vienna 1989, IFAC Proceedings Series, Pergamon Press.

Wybranietz, D., 1990,. Multicast-Kommunikation in verteilten Systemen. Informatik Fachberichte 242. Springer Verlag Berlin, Heidelberg.

METHODS AND TOOLS FOR APPLICATION ORIENTED LANGUAGE (AOL) FOR A COMPUTERIZED RAILWAY INTERLOCKING SYSTEM

P. Axelsson

EB Signal AB, Sweden

Background

The Sternol language is an application oriented language developed for the use in railway safety programs. The Sternol inference engine is designed as diversified programmed software, to fullfil the safety requirements set for railway systems.

The real time and the safety requirements forced us in the the mid 1970:s to define a language of our own, a subset of FORTRAN. This has later (mid 1980:s) been formalized and extended into the current definition of the language STERNOL.

In 1988 a major project to enhance the programmers environment was introduced, bringing the first tools on the SUN work station platform. The goal of the project is to:
- Increase productivity.
- Increase the quality, by easing inspections.
- To enable the safety verification without the target environment.
- To be able to document all tests, and the results, for inspection by safety approving authorities.

The environment project has now (April 1991) reached its first mile stone, with a handfull of tools been used in customer projects for more than 6 month, and with one customer project used as a trial with the new methology.

Properties of the Sternol language

The Sternol language is a declarative language based on multivalue predicate logic (extended Boolean). By using a declarative language the design concepts used in relay based railway equipment can be followed, and thus giving the railway signal engineers the possibility to compare the functions between a relay interlocking and a computerized interlocking.

Our experience is that the use of a declarative language for process control ease the programming task considerably. A formal mathematical completeness test is possible with the use of a language based on mathematical principles like predicate logic. The combination of mathematical proofs, and good methods ensures a deterministic and terminating logic, which is vital in real time process control for safety applications.

The Sternol language uses a strong type check by declaration of the variables. The variables are declared as to value domain, allowed operations by class and by a combination of read/write permissions.

The language constructs are declarative as well. This means that a logical formula is written to define the logical conditions for an assignment of a value to a variable.

The variable binding process is defined by a calculation priority, and variables in a statement during evaluation can not be dynamically bound. This gives a process without true recursion, and is therefore stackless. This is an important feature as stacks not are permitted in our systems out of safety reasons. The processing in an object may be viewed as recursive, but is mearly a means to terminate each instance of the object in order to reduce the communication between instances.

The Sternol language handles the multivalue variables as either a state machine, an enumerated scale, or as positive integers. The different types of variables may be used only in operations that are logical to the type of variable. A code variable (state machine) may for instance only be tested for equality or non-equality, but an enumerated variable may in addition also be tested for relations such as smaller-than, and the integer variables may include the basic arithmetic statements of addition, multiplication, subtraction and division. The possibility to use integers for physical measurements such as meters or seconds has given us the tool to finally after more than 100 years of signalling be able to handle the true problems in the process of railway signalling.

The logic statements are the methods definition in the objects. The methods may be viewed upon as a rulebook for each type of object, such as signals and track switches. Sternol can not be called a true object oriented language, as the language does have only a very limited inheritance mechanism. Still a lot of the object oriented methods and principles are found in the language. Each object is independent of any changes to other objects. Only changes in the global definitions requires a change in all objects. The Sternol inference engine has built in functions to handle communication to and from an operator, to and from the process and between the object instances.

The object instances are created off-line for real time and safety reasons. The object instances contains static information to control the object methods. The static information is vital information to the process, such as the permitted speed through a diverging leg of a track switch. This information is very carefully validated before the commissioning of a system, and must therefore be designed as static information that can not be altered on the target machine for safety reasons.

Design methods
(ref. fig 1)

During the revision of the design methology the major emphasis was put on the test and validation problem, and the possibility to make tools to support the new methodology. The whole concept is based on well known methods and ideas, but with may be a bit of un-orthodox mixing. The main principle is to reduce all design tasks in the the smallest possible sub-tasks that can be tested by its own. Each sub-task is then designed with a mix of the classical "waterfall" method and the object oriented methods.

The basic steps are:

- Drawing and writing a user case description to define the process problems to be solved. The user case description is designed as the system test specification, from the end user point of view, with the idea that if you don't know how to test a function, then you can't design it. When during the later design, questions referring to the process are handled, they are entered into the user case description. The user case description documents are divided by the natural way of dividing the process.

The user case descriptions take the full advantage of the use of a publishing system. We try to use the Donald Duck comic strip as the ideal way of descriptions. Almost everybody can understand the story, regardless of if the reader does understand the language. As most process engineers and software designers have very different languages and concepts of the problems, this has been a very successful way to find a common base for the design.

- The environment description is the next document to be written. This document defines the state machines or algorithms to solve the problem. A major issue in the design phase is to find all possible technical errors that may occur in the input from the process or operator, and how to handle them in a safe way. This document is also drawn and written as a test specification, but for module tests in the development environment. All allowed state transitions are defined, including the conditions for them to occur. The state transitions are written in natural language, but in a format to aid the design and test. All multiple transistions from a state are inspected to check for determinism in the case when all conditions that can possible be valid at the same time is valid. (remember Murhys law?) As a general rule, each state will have one exit transition for normal operation, and one for error handling. The error handling transition is generally given a higher priority in dealing with deterministical problems.

- Diary, minutes of meetings are kept to ensure tracability of even small design decisions from the first meeting with the client to the commissioning of the system.

- Design notes of how the functions will be divided into objects and variables are important for the validation of the design. In addition these notes also reduces the maintenance cost of the code, as it makes it possible to trace the designers thoughts on variable interaction. In most other languages this is written as comments in the code, but in our case we have found the

advantage of using graphics in the descriptions are worth the extra work it is to keep the code and the notes consistent with each other.

- The variable descriptions are a clearified extract from the environment description and the notes, if those documents are written the proper way. The variable description gives a description of the reasons why a variable exists, and a good definition of the value domain. In the case of a state variable each state shall have a description that can be understood even by a newly employed engineer.

- The code design is done in the graphical tool GLE (Graphical Logic Editor), and the code is converted into the publishing system format (still with graphical representation) and all changes are highlighted and explained. This is called the history documents for each variable. The history documents makes it possible to change any release into any other release by doing the changes described in the history documents incrementally. All changes are as a minimum marked with the old release, the new release, designer name, date, and how the change was invoked.
Order number, error report etc.

- The code is tested for the mathematical completeness by the use of the tool CVT (Circuit Verification Tool). CVT test the logical statements for determinism primarily, but also for satisfiability.

- Next the code is module tested off-line in a simulator/debugger, to check that the environment specification is fulfilled. The output is usually post-processed into a graphical format to ease the reading of the output. This is a very efficient way to represent the communication between the object instances. The post-processed trace lists are also in the publishing system format, which makes it possible for us to write comments, highlight and put the test results together with a graphical test specification. The test specification is often neccessary to show how the environment specification items has been tested. (Repeatability.)

In the real life will the smallest possible sub tasks often be possible to divide into even smaller parts as the design continues. Any attempt to finish the specification documents before the release, are doomed to fail. The test of the last smallest function will show missed cases in the user case description. This is not any major problem as long as the this fact is accepted and planned for.

Activity	Tools
Specification	FrameMaker
↓	
Inspection	Wyteboard + Overhead ....
↓	
Composing the descriptions	FrameMaker
↓	
Coding	GLE + FrameMaker
↓	
Validation of the completenes of the logic	CVT + FrameMaker
↓	
Validation of the functionality	EQSIM + EQgraph + FrameMaker
↓	
Compilation	Sternol compiler on a VAX host
↓	
Test in the target environment	TD85
↓	
Release	DOCMAKE +library system

Fig. 1

The tool box
(ref. fig 2)

- GLE + PLOT
GLE is the corner stone in the toolbox. GLE is the only tool used to create and modify the Sternol code. In supplement to the graphical editing facilities, GLE gives support for the variable descriptions by defining the file names, and for automatic consistency check between the source code and the variable descriptions. GLE also includes a rudimentary library support mechanism.

All declarations, both global and local are done by menues, and the logic is designed by creating boxes in the correct "and / or" sequence. The conditions are then written into the boxes. All the normal editing options are included, like copy, cut, paste.

PLOT is a program to create a GLE-style presentation in the publishing system format from the Sternol source code. This is used as a the primary format for all inspections, and design discussions, and also as the base for the history documents.

- CVT + CVTgraph
CVT is a mathematical tool to test predicate logic, adapted to the Sternol multivalue syntax. This is a commercial tool based on a patented algorithm to solve complex logic. CVT basically tests a logical statement for either taotology or satisfiability. The Sternol version has a built in generator to create the proof that only one assignment statement can be true for any input. The proof is created as a predicate logic counter model to the Sternol logic to be tested. As the code sometimes is written with process limits taken into consideration, CVT also accepts a filter statement written by the user, to define conditions that for process reasons can not occur simultaneously.

The query window may in fact be used to write any predicate logic proofs desired.

All output is stored in a file, for future use, or fault analysis.

CVT does also contain a sub tool to find dependencies and inherent possibilities for logical loops. The CVTgraph post-processor presents the dependencies in a graphical format in the publishing system file format.

The full use of CVT requires a good knowledge of the mathematics involved in the form of predicate logic, and the therories involved in solving logical statements. However, the initial cost of the tool and the training is recovered very fast. We have by this tool in a matter of minutes discovered problems which only turns up during very special circumstances, and thus can take years to find. CVT does not produce a correct code, but a stable and robust code.

- EQsim + EQgraph
EQsim is an inference engine designed for debugging and verification of the process. EQsim can handle a set of object instances separate from the Sternol source code. This gives the advantage that a new logic may be tested on the same data as the old logic. The object instances are updated to be consistent with the new object definitions. The processing may be logged on a file, with different levels of tracing.

A Sternol logic package may be used to simulate the actual process input, by reflecting output to the process through the simulation logic. This is a logical system only intended to ease the tests, by simulating the controlled devices.

A great deal of testing is done through command files to EQsim, to ease regression testing. EQgraph is the post-processor for the EQsim log files, creating a graphical presentation of the variable assignments and the messages between the instances.

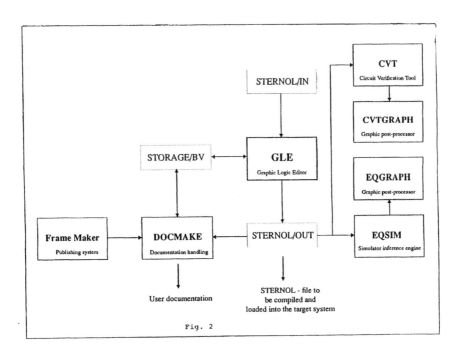

Fig. 2

The EQgraph print outs are compared with the environment descriptions to prove that the function is consistent with the environment description. The tests as defined in the user case descriptions are normally run in EQsim with EQgraph and presented to the customer.

- The Sternol compiler.
The Sternol compiler creates the target machine loadable modules. The compiler is actually two compilers as the logic is used in vital processes. This is the logical consequence of using a diversified inference engine. The Sternol code is processed through the two diversified programmed compilers, and a reverse generation of Sternol is done from the loadables in a diversified way. The reversed code is run through the plot program to enable an inspection of functional equivalance between both of the compiled modules, and the source code. The reversed code is also tested for logical equivalance with the source code by a special comparison function of the CVT tool.

Conclusion
The approval of the Sternol logic for use on-line is the result of a strict review of all documentation related to the system. The safety must be based on sound design from the top, to ensure that the implemented requirements are not a safety hazard by themself. The massive use of graphics is a good investment, as the risk for miss-interpretations is reduced.

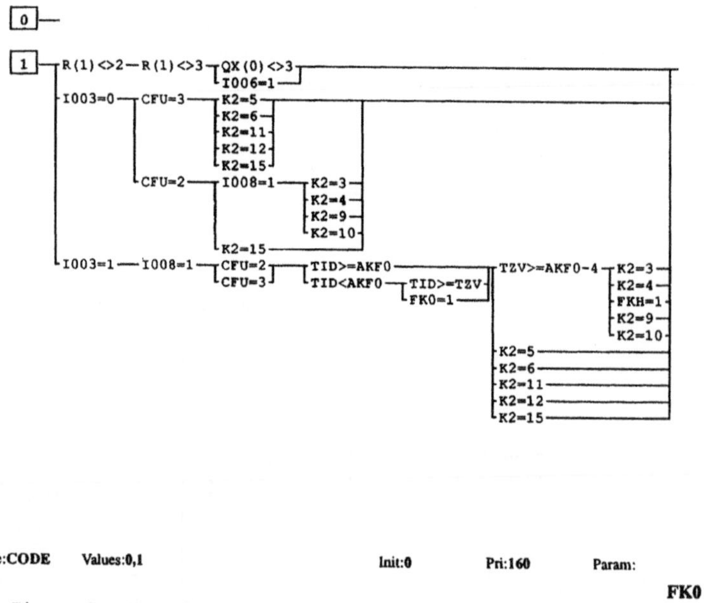

Type:CODE   Values:0,1   Init:0   Pri:160   Param:

FK0

Figure 3   Example of Sternol language

SOFTWARE SAFETY CHECKS USING STORED TESTED PATHS

G. Dahll

OECD Halden Reactor Project, N-1750 Halden, Norway

ABSTRACT

The paper describes an investigation on a technique which idea is to
store all program paths executed during the testing of a safety critical computer program. During on-line execution of this program a check
can be made on whether an actually executed path is found in the file
of tested paths. If an untested path is followed, a special action can
be taken. Various techniques to store the tested paths, and to check
executed paths against the paths in the data base of tested paths were
investigated, with respect to storage requirement, time requirement
and accuracy of approximation. Some of the methods were applicable. A
large number of test runs were made, with different test data profiles, different storing techniques and different faults seeded into
the program. The objective was to investigate the two questions: "how
reliable is the method to trap failures due to residual faults", and
"how often are paths in correct executions not found among the tested
paths", i.e. "how often are spurious actions taken".

Keywords: Software, Safety, Failure detection, Program testing,
Program Paths

INTRODUCTION.

The SAP (Safety Assessment of Software) project
is a joint activity between SRD-AEA technology
and National Power-TEC, both UK, on research
into methods for development and validation of
safety critical software. This project consists
of several activities, one of which is the
activity on Tested Paths. The objective of this
activity is to investigate the applicability of
an idea first put forward by Ehrenberger and
Bologna (Ehrenberger and Bologna,1979) and
(Ehrenberger 1987). The essence of this idea is
to store all program paths executed during the
testing of a safety critical computer program.
During on-line execution of this program a
check can be made on whether the actually executed path is found in the file of tested
paths. If an untested path is followed, a
special action can be taken, as e.g. shut down
or another safety action, a message to the operator or just a logging of the event.

Various aspects of this method were investigated: How to store the tested paths; how to
check executed paths against the paths in the
data base of tested paths; which techniques are
the best for generating the path base as well
as for checking against tested paths during
execution. Another problem which was investigated is how sensitive the method is to the
selection of data used in the generation of the
basis of tested paths.

The basic assumption of the tested path method
is that there is a low probability of a failure
in a path with no failure found at testing.
This assumption was experimentally investigated
by applying the technique on programs with
various known faults.

A final problem to investigate is the frequency
of spurious actions. If a correct, but untested, path is followed, a spurious safety action
will be activated, and if that happens too frequently, the method will not be effective.

METHODS FOR STORING AND CHECKING TESTED PATHS.

The control flow of a computer program module
can be represented as a directed graph of nodes
and edges, and a program path is a sequence of
edges which is traversed during an execution of
the program module. (By a module is meant a
complete program, or a part of it after a
natural subdivision of the program.) A path is
uniquely defined by the set of data used by the
module, i.e. data imported from outside the
module, data defined by initialisation and data
defined by previous executions of the module.
Figure 1 shows a program module with a marked
path.

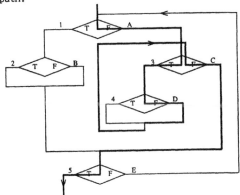

Fig. 1 Program graph with marked path;

Seven different methods were used to generate a data base of tested paths, and to check for tested paths during execution. Two of the methods were exact and five approximate. The exact methods, the BINARY-TREE and the LINKED TREE method used a representation of the set of paths based on a binary tree model. This model assumes that the program module has a single entry (a single exit, however, is not necessary). The path of an execution of the module through the program graph is determined by a sequence of decisions about which edge to follow at each node. Any multiple decision may be considered as a sequence of binary decisions, so that each decision in the path can be represented by a binary decision. A program path is represented by a path trough the tree from the root to a leaf. This is illustrated on Fig. 2.

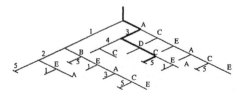

Fig. 2. Binary tree with marked graph.

- The LINKED-TREE method.

In this method the tested paths are represented by a linked set of vertices. Each vertex has two exit branches, each represented by a number. If the branch points to another vertex of the tested path tree, its number identifies this branch. The exit node is indicated with "0", and all vertices of not tested paths have the identification "-1".

- The BINARY-TREE method.

In this method each vertex which is executed in the binary tree is represented by a "1" or a "0", depending on whether one follows the "true" or the "false" branch. The path marked on fig. 2 is thus represented by the binary string 01001. Such bit-strings can be stored very compactly in the computer memory. First of all, n bits can be packed into an n-bit word. A further reduction is obtained by utilising the fact that different paths can follow each other a long way before they split. If the paths are properly ordered, and only the difference between the present and the previous path is stored, a substantial reduction of storage can be achieved.

- The LIMITED-TREE method

This approximate method uses the same storage and checking method as the BINARY-TREE method. Realising that paths which traverse loops many times generate very long paths, this method only stores a limited number of traversals through each loop. In this way the storage requirement is reduced.

- The REMAINDER method

As in the binary-tree method, the paths are stored as binary strings. These string is considered as one long number which is divided with 2147483647, which is the largest prime number less than $2**32$. The remainder from this division is a 32 bits word which is stored as the path identifier. If the bit-string is not longer than 32 bits, the method is exact, otherwise it is approximate. A hashing procedure is used for searching for tested paths.

- The PRIME-operation methods.

The three other approximate methods are based on operations on prime numbers. Each program segment is represented by a unique prime number, and a path is represented by a sequence of operations on these numbers.

- PRIME-MULTIPLY: the numbers are multiplied,

- PRIME-VAROP1: the numbers are operated with the four arithmetic operations +, -, * and / in sequence. (The method used by Ehrenberger and Bologna)

- PRIME-VAROP2: each segment is assigned an arithmetic operation and a prime number, so that the same prime can be used four times.

In the same way as in the REMAINDER method, the paths are stored as an array of numbers, and a hashing procedure is used to search for tested paths.

Program Paths and Module Paths.

The number of distinct paths traversed during real execution easily gets very large in a large program, with long paths and many decision points. It is in practice impossible to make a test which covers all these paths. Instead of storing the paths from entry to exit of the full program (the program paths), an alternative way is to divide the program into modules, and store the paths traversed through the individual modules during testing (the module paths). In this way the number of paths, and also the length of the paths, is reduced.

There is no unique way to divide the program into modules. The simplest is to identify the modules with the subroutines of the program. This is, however, often not the optimum division. A better division would be one in which the number of paths is similar in each module, although that is difficult to know in advance.

Program Instrumentation

To perform the methods described above, an instrumentation of the target program is necessary. The first step is to divide, in a suitable way, the program into a set of numbered modules. The whole program should be considered as one module if one works with program paths.

A subroutines which initiates a path should be inserted at the beginning of each module. Another subroutine is inserted at each decision point and at each loop counter, which adds a new branch to the binary tree, or performs a prime number operation. For the exact methods it is also possible to check whether the executed path is not tested before. At the end of each module a subroutine is inserted which checks whether the path is tested before or not. In the latter case the path can be stored in the path basis. This instrumentation is similar for all the methods, although the checking and storing technique is different.

COMPARISON OF THE METHODS.

To get a first experience with the different methods, they were all applied to a simple test program. The test input data were made with a pseudo-random test data generator. Observations based on the results from this exercise is summarised in table 1.

Table 1 Evaluation of various methods for storing tested paths.

	approxi-mation	storage needed	time for generation	time for checking	check at	arithmetic overflow
LINKED-TREE	exact	very large	good	good (best)	each node	no
BINARY-TREE	exact	accept-able	(very) large	accept-able	each node	no
LIMITED-TREE	poor	accept-able <BIN-TRE	large	accept-able	each node	no
REMAINDER	good	good	good	good	end only	no
PRIME-MULTIPLY	poor	good	good	good	end only	probable
PRIME-VAROP1	good	good	good	good	end only	possible
PRIME-VAROP2	accept-able	good	good	good	end only	less probable than VAROP1

The two exact methods have both implementation drawbacks, viz. the storage requirement for the LINKED-TREE and the generation time for the BINARY-TREE method. The storage requirement for the LINKED-TREE method may be very large if the paths are long. This method is therefore only applicable for module paths, since such paths are shorter and fewer than the paths of the complete program. The large generation time for the BINARY-TREE method can be accepted, since the generation is made during testing only, and in addition the generation procedure probably could have been made faster by optimum programming. However, the checking time for the LINKED-TREE method is faster, so if one had enough storage capacity, this method would probably be the best choice of the two.

The REMAINDER method seems to be clearly the best among the approximate methods. It is faster, simpler and more accurate than the other approximate methods, so these were not investigated any further in the project.

APPLICATIONS TO REALISTIC EXAMPLES.

The further investigations of the tested path method was made with more realistic programs, viz. the two trip programs TRIPC and TRIPV previously developed in the PODS project (Bishop et. al. 1986)

Test Data
A set of test data generators were also developed in the PODS project comprising the following types of data:

- Systematic data
  This is a set of 3144 input data sets selected to cover all aspects of the specification, i.e. combinations of data from all the different ranges of all input parameters.

- Random data
  Input data generated during testing with the help of a pseudo-random number generator. Generators were made to different data profiles: uniform random, Gaussian and distributions round the borders in the input data domain.

- Plant simulation data
  A simulator was developed to produce sequences of data which reflects a realistic operation of a plant. This consists of start-up, normal operation with small random changes and shut down. In addition a pseudo-random

generator is used to inflict transients, faulty input data, recalibration etc. Parameters can be used to change the frequency of such events.

Accuracy of the REMAINDER Method.

An investigate was made of the accuracy of the REMAINDER method, i.e. how often two distinct paths come up with the same remainder, and are therefore indistinguishable in the tested path base. This can be measured by counting the number of distinct paths this method finds and compare it to the number found by the exact BINARY-TREE method. A large number of test runs were made, and the result showed that the approximation was very good (on average about 1 in 1000 paths were double counted).

The REMAINDER method is a very effective method to store and check for tested paths, and was therefore used for the further investigations. The main drawback with this method is that it can only check paths at the module exits, and will therefore not be able to trap infinite loop failures. The exact methods can check at each decision point whether the executed path starts to deviate from any tested path.

THE EFFECTIVENESS OF THE TESTED PATH METHOD.

The basic philosophy behind the tested path method is that less failures should occur on a properly tested and verified path. A question to investigate is therefore: "will the tested path method really reveal the faults", or alternatively: "are tested paths more reliable than untested ones".

Another key question concerns the possibility that a special safety action is activated for correct (but untested) paths. In other words: "what is the probability of spurious activation?" This probability should not be too high for the method to be effective.

To investigate these questions a set of experimental test runs were made. The method used was to seed faults into the two trip programs TRIPC and TRIPV, and then to run each of them back-to-back with a "golden" version. A set of 30 faults in TRIPC and 22 faults in TRIPV identified in the PODS and STEM (Bishop et. al. 1987) projects were seeded back in to generate 52 bugged programs.

Reliability of Tested Paths.

For each of the selected faults the following procedure was followed: A correct and a "bugged" program were executed back-to-back with up to 5000 cycles of input data to create a set of 1000 cycles where there is no discrepancy on the outputs of the two programs. Faults with less than 10 failures or more than 4000 failures were discarded (i.e. only faults with failure probability in the range <0.002,0.8> were examined further). When a discrepancy occurred, the corresponding cycle number was stored in a file. In this way a test data set with 1000 cycles of test data is generated. This data set generates no faults, and thus "misses" the seeded fault. This simulates an incomplete test.

The next step was to investigate how well the tested path method is able to trap the faults that were not found by the above mentioned set of test data. A path basis was generated from the execution of a program with these 1000 cycles of test data. Then the correct vs. the bugged program was executed 5000 cycles and the four parameters: number of failed/correct cycles with tested/untested path were measured.

The optimum result would have been if the number of failure cycles with tested path was zero, i.e. the paths of all failed cycles were outside the tested path base, and a safety action would be activated. This does, however, not agree with the computed results. A question is therefore how good the results are, i.e. what do the results show on the reliability of the tested paths.

One way to investigate this question is to compute the probability that no safety action will be taken at failure. The parameter:

$$prnaf = \frac{\text{number of failed cycles with tested path}}{\text{number of failure cycles}}$$

is a measure of this probability which can be computed on the basis of these results.

If the probability of hitting a tested path is small for all cycles f.ex. if the path basis is small, prnaf would also be small. This effect would bias the value of prnaf, and to compensate this biasing, this probability should be weighted with the probability of not executing a tested path at all. A relevant parameter to measure the reliability of the tested paths is thus:

$$na\_ratio = \frac{prnaf}{\text{probability of no action at all}}$$

and the result is shown in table 2.

A value close to zero indicates a positive effect, whereas a value near is neutral. A value greater than unity indicates a negative effect. The latter is difficult to explain, but may be due to statistical uncertainty.

Investigation of "Pure" and "Mixed" Paths.

The terms "pure" and "mixed" paths will be used with the following meaning: A path is "pure" if all executions of the path are correct or if all executions fail. The path is "mixed" if some executions of the path are correct and some executions are faulty. The "purity" of a faulty program is low if there are many "mixed" paths during execution.

The tested path method should work best with a high "purity" of the executed paths. To investigate the validity of this assumption back-to-back runs were made with a correct program

TABLE 2 Reliability of Tested Paths.

The table shows the value of na_ratio for different bugs in TRIPC and TRIPV.

Uniform random data

TRIPC fault	progr. path	module path		TRIPV fault	progr. path	module path
3	0.196	0.876		1	0.000	0.470
4	0.000	0.000		2	0.000	0.000
5	0.000	0.983		3	0.000	0.413
6	1.211	0.003		4	0.738	0.919
7	1.638	1.015		6	0.000	0.000
9	0.000	0.000		11	0.939	0.857
10	0.191	1.041		12	0.000	0.904
11	0.000	0.000		15	0.185	0.870
12	1.106	1.029		16	0.000	0.965
13	0.916	0.916		17	0.085	0.071
15	0.000	0.961		18	1.311	0.953
16	1.022	0.999		19	0.000	0.978
18	0.816	0.489		21	1.222	0.536
19	0.495	0.915		22	0.427	0.911
21	0.041	0.897				
22	0.000	0.658				
23	0.000	0.000				
26	2.052	1.003				
29	0.359	1.006				

Plant simulation data

TRIPC fault	progr. path	module path		TRIPV fault	progr. path	module path
4	0.125	0.000		1	0.000	0.000
5	0.000	0.997		2	0.000	0.000
6	0.000	0.000		4	0.599	0.852
10	1.500	1.821		11	0.968	0.990
12	0.571	0.789		13	0.000	0.000
13	0.005	0.775		18	0.000	0.000
17	0.000	0.000				
29	0.693	0.883				

TABLE 3. Mixed Path Computations
The number of distinct mixed paths (npmix) and the probability of failure at mixed paths (pfamix) for different faults.

TRIPC fault	npmix	pfamix		TRIPV fault	npmix	pfamix
2	13	0.9262		1	0	--
3	52	0.8021		2	0	--
4	23	0.1399		3	0	--
5	0	--		4	90	0.1813
6	0	--		6	0	--
8	0	--		7	13	0.1118
9	0	--		11	213	0.5037
10	26	0.0458		13	0	--
11	0	--		15	5	0.1754
12	55	0.1346		16	0	--
13	15	0.7857		18	0	--
15	17	0.1490		19	51	0.0527
17	0	--		21	18	0.4811
18	23	0.8942		22	56	0.9112
19	21	0.9398				
21	24	0.7195				
22	14	0.1902				
23	0	--				
27	9	0.2691				
29	96	0.0894				
30	1	0.6667				

versus all the bugged ones was made with 5000 cycles of plant simulation data. In each of 5000 executed cycles of plant simulation data, the traversed path was identified and logged together with information on whether the execution was correct or failed. The number of distinct pure and mixed paths, as well as the number of execution cycles in each of these categories was measured.

The optimum result is obtained if there are no mixed paths. If this is not the case, a relevant measure is the probability of failure at mixed paths which was also computed. The results are shown in table 3.

Spurious Activations.

If the number of paths during execution is large and the number of tested paths is relatively small, it is fairly probable that an executed path is not found in the tested path base, but the execution is still correct. In this case a spurious safety action will be activated, and if that happens too frequently, the method is not very useful. The most relevant parameter to use as an indicator is the probability of spurious actions, see table 4.

$$Pspac = \frac{\text{no. of correct cycles with untested paths}}{\text{total number of correct cycles}}$$

TABLE 4 Values of Pspac for Different Bugs.

Uniform random data

TRIPC			TRIPV		
fault	progr. path	module path	fault	progr. path	module path
3	0.808	0.039	1	0.996	0.069
4	0.902	0.037	2	0.997	0.058
5	0.864	0.035	3	0.997	0.071
6	1.000	0.036	4	0.997	0.069
7	0.904	0.036	6	0.998	0.079
9	0.901	0.037	11	0.997	0.065
10	0.906	0.040	12	0.997	0.081
11	0.864	0.038	15	0.996	0.070
12	0.905	0.039	16	0.997	0.072
13	0.918	0.046	17	0.991	0.034
15	0.891	0.039	18	0.998	0.077
16	0.898	0.038	19	0.998	0.088
18	0.894	0.035	21	0.998	0.080
19	0.816	0.036	22	0.996	0.058
21	0.831	0.042			
22	0.878	0.037			
23	0.867	0.032			
26	0.926	0.039			
29	0.879	0.038			

Plant simulation data

TRIPC			TRIPV		
fault	progr. path	module path	fault	progr. path	module path
4	0.265	0.047	1	0.303	0.127
5	0.307	0.047	2	0.281	0.100
6	0.251	0.073	4	0.225	0.102
10	0.505	0.482	11	0.188	0.040
12	0.415	0.173	13	0.093	0.015
13	0.393	0.158	18	0.292	0.120
17	0.197	0.010			
29	0.416	0.151			

Interpretation of the Results.

The objective of this exercise was to find out how effective the tested path method is as a technique for on-line safety checking. There is no definite answer to this since the results vary for the different types of faults. The performance would be maximised if the number of failures in tested paths is zero, the number of correct cycles with untested paths is zero and there are no mixed paths. These ideal results have not been reached. Some questions to answer are then: How much deviation from the ideal result should be allowed in order to state that the result is positive? Which are the most relevant parameters to use in the evaluation of the result? How should the difference in the various tables be understood? The various aspects of tested path performance are discussed in the following:

- Reliability of tested paths.

The tables show a large variation in the results. Particularly from fault to fault, but also between test profiles (random and simulation) and between path storage method (program paths and modular paths).

An observation is that the ability of the tested path method to reveal hidden faults not found during testing depends on the "anatomy" of the fault. Some faults are "path dependent", i.e. they will always make a failure if the paths they belong to are traversed. Such faults should be trapped by the method. The other type is "data" dependent. This means that the program may, or may not, fail on the faulty paths, depending on the value of certain data. Such faults would not necessary be found by the method, but might be better handled using a technique checking for tested data.

TABLE 5. Mean Values of na_ratio and Pspac.

prog.	data type	path type	na_ratio	Pspaxc
TRIPC	uniform rand.	program	0.529	0.887
TRIPC	uniform rand.	modular	0.673	0.038
TRIPC	plant simul.	program	0.362	0.344
TRIPC	plant simul.	modular	0.658	0.143
TRIPV	uniform rand.	program	0.350	0.997
TRIPV	uniform rand.	modular	0.632	0.069
TRIPV	plant simul.	program	0.261	0.231
TRIPV	plant simul.	modular	0.307	0.084

To check for any differences between data profiles and path storage methods, the mean values of the ratios were computed for all combinations (see table 5). The ratios are smaller for simulation than for random data, and also smaller for program paths than for for modular paths, but not essentially smaller. These results support the conclusion that modular path storage is the better strategy. On average, the ratios are below unity (0.2 - 0.7), but the improvement with the method is only modest.

The results could be biased by the lack of realism in the experiment, since:

- Some of the faults studied had very high failure rates, these could give unrepresentative results.

- In some tested cases, the number of executions of either the tested untested paths was very low, so that the failure probability estimates are not very exact. Longer test runs are needed to obtain precise results.

- The path basis may be unrealistic, especially for the high probability faults. As designed, the test procedure inserts mixed paths into the path basis provided one correct path execution is made. An alternative approach is to exclude all mixed paths from the path basis. This should result in smaller ratios, since only 'pure' correct paths are included in the basis.

- An alternative view is that, in practice, tests would be applied until a failure occurs. This puts an upper bound to the number of tests, and hence the number of tested paths, that can be applied while leaving the fault intact. In general the maximum would be the inverse of the failure probability of the fault under the test conditions. For many of the present faults the high failure probability would lead to a very small number of tests before fault detection. The only way to obtain a reasonably large test basis would be to use much smaller faults.

Even though questions may be raised over the realism of the experiment, the overall result may still be realistically correct.

- *Mixed path probability.*

The computation of the number of mixed paths and the failure probabilities in mixed paths are good complementary indicators to evaluate the tested path method. They can also be computed for some faults which were skipped in the first type of measurements. It is not applicable when using uniform random data, since these data seldom traverses the same path more than once. (It could, however, have been applied to the module based paths.)

The various program faults exhibit different behaviour. Some faults have no mixed paths. Of the remaining faults, some have a combination of 'pure' failure paths and mixed paths, while the rest are almost entirely composed of mixed paths. This behaviour seems to depend on whether the faults are 'path-dependent' or 'data-dependent'. It is for the latter type of faults the mixed paths occurs.

- *Spurious activations.*

The theoretical maximum of program paths is very large. In principle this could result in very high spurious activation rates, since there is a very large population of untested paths. The assumption for the tested path method is, however, that the number of executed paths during real operation is much more limited.

To check for any differences between data profiles and path storage methods, the mean values of the ratios were computed for all combinations, in the same way as for the ratios. The results are also shown in table 5. The probability of spurious activation is clearly less for modular paths than for program paths, and also less for simulation than for random data.

It is, however, possible to reduce the spurious activation rate, by

- Applying more tests to increase the size of the path basis.

- Using module-based paths, where the theoretical maximum of paths can be reduced drastically compared to program paths.

- Improving the coverage by recording the paths traversed during module acceptance testing.

- Using real process data if they are available. This would ensure that the most frequently executed paths during real operation are in the tested path base.

- Combining the paths from the module acceptance and process data tests with data from "accelerated" plant simulation tests. In this way a good coverage of operational paths should be achievable.

CONCLUSIONS.

Methods for simple identification and storage of paths executed during testing has been developed and investigated. It was found that: The approximate REMAINDER method is simple fast and accurate, and requires the smallest amount of extra storage. Of the two exact methods, the BINARY-TREE method is acceptable concerning extra storage and checking time, but more complicated than the other methods, whereas the LINKED-TREE method is simple and fast, but can only be applied to module paths where the paths are short and not too numerous. Only the exact methods can trap failures causing non-termination.

These methods can be useful for on-line checking, as originally intended with the project. But the methodology can also be used as a measurement during testing. Number of paths can be used as a measure on test completeness. Number of distinct paths per test data input cycle is a measure of the effectiveness of the test data. And the overlap of the path set generated by the test data compared to a path set generated by real process data may be seen as a indicator of the relevance of the test data set.

Spurious activations of the special safety action of the tested path method constitutes a particular problem. This is due to the incompleteness of the set of tested paths. The best test data for path generation would be a large set of real process data, to prevent too many spurious activations, plus a set of random data to catch subtle faults.

The best way to reduce the problem of spurious activations is to divide the program into modules, and use module based paths.

Only certain types of faults can be revealed with the tested path method. Some faults are "data dependent" rather than "path dependent" and a checking against a basis of tested data may also be needed.

A final conclusion is that path coverage is not a sufficient measure for a complete testing.

REFERENCES

Bishop (1986) Bishop et al. "PODS - A Project on Diverse Software". IEEE Transactions on Software Engineering Vol SE-12, Number 9, 1986

Bishop (1987) Bishop et al. "STEM - A Project on Software Test and Evaluation Methods" in "Achieving Safety and Reliability with Computer Systems" edited by B. K. Daniels, Elsevier Applied Science, 1987

Ehrenberger and Bologna (1979) W. Ehrenberger and S. Bologna "Safety Program Validation by means of Control Checking", Proceedings from SAFECOMP'79. Pergamon Press 1979.

Ehrenberger (1987) W. Ehrenberger "Fail-Safe Software - Some Principles and a Case Study" in "Achieving Safety and Reliability with Computer Systems" edited by B. K. Daniels, Elsevier Applied Science, 1987

ENHANCING SAFETY BY DIVERSITY - ONE MORE WAY TO IMPLEMENT IT

B. Sjöbergh

Technical Dept., AT Signal System AB, Spånga, Sweden

Abstract. Today, micro-processors are used in many safety systems. One of the main concerns of such systems is the effect of software on system safety. Software diversity, that is, more than one program, is a common solution. This has effects on economy, availability and maintainability as well as on safety. We have implemented a slightly different solution where the second program is only used for the validation and testing of the produced software. The advantages are ease of modification and high system availability at reduced cost. Different types of error sources and error types and how to find and avoid them are discussed.

Keywords. Software engineering; safety; validation and testing;

INTRODUCTION

Hardware safety systems can usually be designed and analysed to be safe. As system complexity increases hardware solutions are less likely to be practical and cost-effective. Instead, computer-based systems are the natural solution. These can also be designed to be safe as the hardware systems, but only as far as hardware failures are concerned. The complication is that many functions are software-based and software can seldom be proven to be safe.

As software is used to implement complex functions, software is usually complex, even at its best. It is difficult to test it and understand it completely. To understand the behaviour of software under fault conditions is also very difficult.

To overcome the safety problems a two-program solution is often chosen. This is done on the assumption that the errors in different programs are not correlated. To enhance safety, this must be proven for the actual system. It must be remembered that two wrongly designed and/or implemented units does not automatically make one correct safety system! Systems can contain errors of several types coming from different sources. These must be handled in a way appropriate to each error type.

ERROR TYPES

When a software-based system is used, two kinds of failures can occur. Of cause components can fail and noise can cause errors, just like a hardware-only system. But errors can also occur repetitively because of software bugs. Bugs come from lack of knowledge, misunderstandings, misinterpretations, carelessness etc. They can find their way into customer specifications, system designs, low-level specifications, source code, machine-code etc.

Customer Specifications.

Errors at this stage can cause serious trouble and be very expensive and time-consuming to correct. It is most important to ensure that this spec is well understood by both parties and that it represents the true requirements of the customer organization, not just that of a few individuals.

System Design

This stage can also cause devastating error effects. It should be checked thoroughly by several separate teams (analysis diversity).

Errors here can be both misunderstanding the customer requirements and process, faulty design and lack of specification, leaving too much room for errors later on. It is also possible to overspecify, making implementation difficult.

Low-level And Interface Specifications

These specs have far-reaching effects and should be well analysed. When using diverse software, more than one group automatically check them. This should not be regarded as a complete analysis, though.

As these specs are normally common to all programs, diversity programming does not find errors at this level.

Source Code

At this stage, errors have less far-reaching effects but can still cause dangerous situations. They can easily be corrected when found. Testing can find this kind of errors but can be difficult to do exhaustively. Analysis is a powerful tool for error removal. Code walk-throughs can also be used but here personalities can convince more than the actual code.

Coding diversity can detect these errors. If the programs are very similar, this method is inefficient. That means that it is probably an advantage to use different coding methods, strategies and languages.

Machine Code

These errors are systematic, caused by compiler bugs etc, or random, caused by noise or failures. They usually have local effects only. Testing is normally the only way to find them.

The random errors are easy to correct, but the systematic can be difficult. Either a way around them has to be found, or tools must be changed. Changing compiler or release can easily bring new bugs and so on.

Coding diversity can detect these errors if used correctly. The systematic errors can only be detected if different tools such as compilers or even languages are used.

Run-time Only Errors

Noise and hardware failures can cause strange software behaviour. The way to combat this is by hardware diversity and run-time self-tests.

SOFTWARE DIVERSITY PROBLEMS

For diversity to eliminate common errors, programs must be very different and still conform to the same specification. This is not necessarily the case, as reported by Knight and Leveson (1985, 1990); within an organization common ways of programming and common tools make the risk of common errors very real. Intermediate result comparison methods also mould the programs into similar form, lessening diversity. Another interesting text on programming and analysis diversity is N-fold inspection by J Martin and W T Tsai (1990).

Diversity can also cause availability problems, if minor differences cause frequent shut-down of the system. A single program seldom disagrees with itself!

Having to write two programs and make them run well together to be able to test new functions is awkward and expensive. Keeping two or more programming teams for program maintenance is also costly. To keep the teams isolated from each other for years is difficult.

ATSS ATC2 - A PARTIAL SOFTWARE DIVERSITY SYSTEM

The ATC2 software is designed to enhance the existing ATC system in use in Sweden and Norway for automatic train protection. In short, it ensures that the train driver does not exceed the maximum allowed speed from signals and speed-boards.

The new software is based on the same design as the present system software. It is a cyclic scheduling system with two interrupt levels for timing and high-speed data

input. In use, it is a hardware-redundant two-out-of-three system with a single program. This has made the system a good test-bed for modifications to the customer requirements specification.

During verification and validation, a second program is used. This program has been written in ADA, a very strict high-level language designed for US DOD. As the first program is written in assembly language, the two are very different. The two programs are compared on an automated test-rig running on a PC.

The next paragraphs explain how the different error categories have been handled.

Customer Specification Errors

A lot of people from the customer and two equipment suppliers have been involved in generating the spec. Details of the changes have also been tested in real-life with special test-versions of the software (non-validated).

This has also ensured good mutual understanding of the problem and the spec.

System Design

The old design that has given good service for ten years has been re-used. This design is fairly simple and well known and has been analysed by several people over the years.

Intermediate And Interface Specifications

These have been checked by the ADA programming team as well as by a separate software analyst.

Source Code

This has been tested against the ADA implementation module by module to find coding and documentation errors. It has also been analysed thoroughly to ensure good and stable design, adherence to specifications and correctness.

Machine Code

This has been tested against the ADA implementation.

Run-time Only Errors

These are taken care of by the two-out-of-three design.

System Tests

To detect any remaining errors the system has been tested on a simulator as well as on a railway line in Sweden equipped for this kind of test. These tests aim at checking the unspecified details of real-life surroundings as well as the system-level implementation and interpretation of the specification.

CONCLUSION

The partial software diversity method as described here has shown good results: it has not been the only

validation method, not even the most important, but still
a useful tool. It has been complemented by much
analysis and testing to ensure adequate system safety
levels. In this respect it is no different from a complete
diversity implementation.

The diversity programming has mainly affected the
program by implying and checking the adherence to very
strict high-level-language rules for the PDL
specifications and documentation of the assembly
language program and its variables. If only assembly-
language is used, these aspects easily gets into a sad
state, making system maintenance very painful and
error-prone.

REFERENCES

Knight,J.C., and Leveson, N.G. (1985). An experimental
evaluation of the assumption of independence in
multi-version programming. IEEE Transactions on
Software Engineering vol SE-12 no1, 96-109.

Knight,J.C., and Leveson, N.G. (1990). A reply to the
criticisms of the Knight and Leveson experiment.
ACM Sigsoft Software engineering notes vol 15 no
1, 24-35.

Martin, J. and Tsai, W.T. (1990). N-fold inspection: A
requirements analysis technique. Communications
of the ACM vol 33 no 2, 225-232.

SYSTEMATIC SOFTWARE TESTING STRATEGIES AS EXPLANATORY VARIABLES OF PROPORTIONAL HAZARDS

F. Saglietti

Gesellschaft für Reaktorsicherheit (GRS) mbH, Forschungsgelände, D-8046 Garching, Germany

Abstract. The use of Proportional Hazard Models (PHMs) to characterize different software environments is discussed: this is done in particular by comparing the major characteristics of systematic testing and operational settings. The intention hereby is to draw conclusions from test failure data to operational reliability. After a formal definition of the modelling class considered, particular testing strategies are identified, which may be suitably modelled as explanatory variables of PHMs. The theoretical concepts introduced are exemplified by real-world data. The conclusions include an evaluation of test diversity as well as a choice optimization among alternative testing strategies.

Keywords. Computer software; software engineering; program testing; error analysis; failure detection; reliability theory; probability; modeling.

Acknowledgement. The investigations reported have been sponsored in part by the Commission of the European Communities under the ESPRIT program (subitem Software Technology, projects REQUEST and DARTS): the author thanks for the support.

INTRODUCTION

The present study is the result of considerations on the use of *Proportional Hazard Models* (PHMs) to characterize differing software environments in terms of their operational reliability.

Approaches based on PHM-theory are known to assume the failure rate being decomposable into the product of a base-line hazard and of an exponential term incorporating the effects of particular explanatory variables.

Several applications of this theory have already been proposed in the past in a number of reliability areas, e.g. in the *medical field* and in *stress testing* of physical devices. They are intended to study the impact of specific environmental characteristics, as human life conditions or machine overload factors, on the expected failure probability with respect to particular fault classes: in the former case specific diseases, in the latter example deteriorations of single components.

For software dependability evaluations an analogous approach could be applied to compare different *usage patterns* of the same product and to predict, on the basis of old failure observations related to a given profile, the new user-perceived reliability before actual operation in an unexplored environment (Littlewood (1987/88)).

In particular, such a model of the diversity of operational settings would also be extremely useful in order to draw conclusions from test failure data to operational reliability by taking into account the differing characteristics of *testing and operational environments*.

This is the main intention of the following investigations: they aim at identifying particular testing strategies, which can be suitably modelled as explanatory variables of PHMs: thus we may compare system operational behaviour before and after a systematic testing phase by evaluating the proportionality effect of the specific strategy considered on the expected hazard curve.

The next chapter will be devoted to a formal definition of Proportional Hazard Modelling, leading to some considerations on its general applicability.

PROPORTIONAL HAZARD MODELLING

The dependability of system behaviour is usually evaluated in terms of the underlying process hazard rate $\lambda(t)$, given by defining, for an infinitely small interval Δt, the product $\lambda(t) \cdot \Delta t$ to be the probability of almost immediate failure of a component of age t.

Some of the existing software reliability growth models provide a continuous approximation of the hazard function of a software product on the basis of past failure data obtained during execution time by selecting the inputs according to an operational profile P.

According to PHM-theory, the hazard function is assumed to be decomposable into the product of a *base-line hazard* $\lambda_0(t)$ and of an *exponential term* incorporating the effects of the *explanatory variables* $z_1, ..., z_k$:

$$\lambda(t) = \lambda_0(t) \cdot \exp(\beta_1 z_1 + ... + \beta_k z_k),$$

$$\text{with } \beta_i \in R.$$

Although "the assumption of a proportional effect is the only assumption that the technique makes" (s. Wightman and Bendell (1986)), it certainly represents a very heavy restriction of generality, which should not be underestimated by considering PHM as a suitable cure-all for treating the impact of any explanatory variable.

In fact, for two arbitrary environments given by different operational or testing input profiles, there is no reason to expect proportionality between the respective hazard functions; in general not even a monotonicity relation will be granted, as different input selections will detect failures in non-comparable order and frequency.

In the past, different experiences were reported about the dependence of failure distribution on the type of system utilization. Mourad and Andrews (1987), for example, compared two identical IBM systems w.r.t. their error occurrence, which is highly determined by the different operational activities. Rossetti and Iyer (1982) related system reliability to the usage environment by means of quantifiable workload variables.

In close analogy to the classical hazard concept, they represent the *apparent load hazard* (a measure of the incremental risk involved in increasing the workload from x to x+Δx) as the product of a *fundamental hazard* (an inherent property of a particular system, not subject to varying load patterns) with a probabilistic workload measure.

These experimental results confirm that the correlation between failure behaviour and operational usage of a system is by far more complicated than the simplistic proportionality hypothesis of PH-models would imply.

Nonetheless, there may be particular occasions, where the very peculiar PH-assumption formulated above may be regarded as being fulfilled. Such a very special case will be considered throughout this paper.

UNIFORM TESTING STRATEGIES

According to the input distribution given by the selection probability P, the errors are detected in the same order as their operational seriousness (s. Currit, Dyer and Mills (1986)). This naturally results into intervals between failures becoming larger with time, which may render too arduous the problem of achieving and assessing a given minimal reliability requirement, as the necessary test effort may grow beyond any realistic possibility.

This is certainly one of the main reasons why data provided during such a *statistical* testing phase are integrated with further information coming from so-called *systematic* test strategies. Systematic testing is hereby defined to represent all strategies selecting the input x not just according to a predefined "black-box" probability P(x), but taking into account the *functional* or *structural* program features which will be exercised by executing x. A comparison and possible combination of systematic and random input selections was proposed by Saglietti (1988).

Systematic testing phases are therefore capable of finding errors regardless of their "failure size". A very special case among them occurs under the additional condition that each error is (approximately) equally probable to be detected. We will call a test with such a property a "*uniform test*". Of course, there is no evidence for this peculiar characteristic; at best it may be assumed with respect to some specific error classes.

The ideal case is represented by an input space partitioned into subsets such that within each subset affected by an error, there is the same probability of detecting it. For instance, If we assume each error to affect all of a path (or the same path proportion), then uniform path testing (e.g. s. Puhr and Krzycacz (1982)) may be regarded as an example for a uniform test.

For an arbitrary systematic strategy as a functional or structural test this assumption may possibly be not exactly fulfilled; nonetheless, we may regard it also as an expression of subjective indifference towards the order in which errors are detected.

In fact, even if the single detection probabilities are different, we may suppose that their ascending order is not necessarily related to the corresponding error size order, each detection probability being essentially determined by the ratio of *error size* to *partition size*.

In this case we may regard the probability of detecting an error during a systematic test as independent from the time of its occurrence during an operational phase, so that we may consider throughout execution time an *average value* (instead of an exactly constant one) for the probability of systematic error detection.

This extension of the restricting assumption may allow to regard the results achieved under exact conditions in the following paragraphs as possible approximations of reality, intending to provide a means to estimate the average impact of any systematic testing phase on the operational failure behaviour.

As the approach suggested is essentially based on comparing *random failure detection* with *random error detection*, we have to point out that the underlying correspondence error ↔ failure is not always well-defined. While this relation is unequivocal in case of disjoint failure subsets, if their intersection is non-empty the error detection and correction phase may allow different interpretations.

For example, we may consider 2 intersecting failure sets caused by 2 errors also as the union of 3 disjoint failure sets caused by 3 different errors.

Of course, the actually realistic interpretations will heavily depend on the error classes and on their impact on the debugging phase. Anyway, this observation represents a restriction of our theory, as it limits the generality of the term "uniform test" being well-defined. Nonetheless, if we assume reasonably reliable systems with most failures being disjoint or optimally corrected, we may regard this weakness as negligible.

The next section will be devoted to proving the proportionality of the hazard functions of the same software having been resp. having not been subjected to a uniform test. The conclusion is that such a systematic phase can be suitably modelled as explanatory variable by means of a PHM.

THE IMPACT OF A UNIFORM TEST ON THE HAZARD FUNCTION

In this section we will study the impact of a uniform test T on the hazard function of a program P_0. In other words, denoting by P_0 the original yet untested program and by P_T the version resulting from P_0 after the testing phase T, we will compare the hazard functions $\lambda_0(t)$ resp. $\lambda_T(t)$ of both software products with respect to the same operational profile, according to the picture shown in Fig. 1.

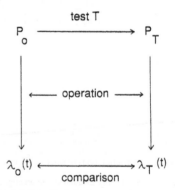

Fig. 1.   Relations between programs and hazard functions considered

Being a uniform test, T will discover each error with the same probability, say C_T. The probability given by $\lambda_T(t) \cdot \Delta t$ that the tested version P_T of operational age t will fail in the next instant is then exactly the probability of the original program P_0 of same age failing in the next moment and of the error responsible for this failure not having been discovered by test T, i.e.:

$$\lambda_T(t) \cdot \Delta t = \lambda_0(t) \cdot \Delta t \cdot (1 - C_T), \text{ i.e.:}$$

$$\lambda_T(t) = \lambda_0(t) \cdot U_T$$

$$\text{with } U_T := 1 - C_T$$

In particular, this intuitive argumentation immediately yields the proportionality of hazard functions required to apply a PHM.

The main result of this proof is the possibility of establishing a connection between the program improvements achieved by a *statistical test* (possibly performed during simulated operation) and by a *systematic testing* phase intended to cover structural properties or functional requirements.

The applicability of such a PHM-approach has the advantage of permitting both types of testing strategies to be carried out in any arbitrary order: the systematic test may be previously performed and then integrated into an operational phase, but it may also become necessary to insert such a selective phase as the failure rate resulting from statistical testing may be too high ($\gg \lambda_{max}$) for product validation, but still too flat to promise the desired decrease within a realistically affordable testing duration, as sketched in Fig. 2.

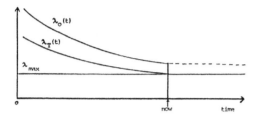

Fig. 2. Hazard rates resulting from statistical and systematic tests

In this case one may evaluate the resulting curve $\lambda_1(t)$ obtained by stretching the original curve $\lambda_0(t)$ by a factor C_T. This could be possibly done by comparing the failures occurred during random input selection with those detected by the alternative systematic strategy T, estimating C_T from the past error detection capability of T.

Let E_S denote the set of errors detected by the statistical test, and E_T denote the subset of such errors, which were also detected during the systematic testing phase. In case of sufficient observations we may obtain a first rough approximation of checking efficiency by the following quotient:

$$\Rightarrow C_T \cong |E_T| / |E_S|.$$

This evaluation method clearly reminds of the well-known procedure called "*seeding and tagging*", which was originally used by Feller (1957) to estimate unknown animal populations. It mainly consists of capturing, releasing and recapturing (by identical means) samples of animals, extrapolating on their total number on the basis of a comparison of both catches and their intersection.

In our case we also apply a replicated "*capture-recapture*" procedure in order to compare both processes. The main difference, however, lies in our dissimilar methods of capturing and of recapturing: statistical testing by operational simulation may be compared with a fishing-net of varying mesh size, so that bigger exemplars are caught with higher probability than smaller ones, which may be let through; uniform testing, on the other hand, rather corresponds to a fish-hook with a non-specific bait, attracting all animals of a species regardless of their size.

Due to this fundamental variance in the techniques used, we cannot simply apply the hypergeometric distribution to evaluate testing efficiency, as usually done by modelling mutation tests.

Significant estimations of C_T can be obtained, however, by taking into account the specific features on which the testing strategy T is based. As already mentioned, the systematic criteria may themselves include randomness; this is, however, strictly related to a predefined *input partition*, which is *deterministically* fixed by the program properties to be checked.

The *probabilistic* aspects of a systematic test are rather concerned with the *selection strategy* of the partitioned subsets, which may be determined by one or more of the following parameters:

n1 = total number of subset selections,

n2 = number of different subsets selected,

n3 = total number of test cases out of each subset selected.

Anyway, first the subsets (and then within the selected ones the inputs) are assumed to be selected with the same probability, as shown in the following examples.

Let N denote the number of input subsets. In the following we will consider the class of errors affecting each subset with identical proportion. More precisely, we assume each error to have a constant probability θ of being detected by *one* test case out of the corresponding subset. Then we can distinguish the following uniform tests and easily evaluate their constant detectability C_T per error:

a) *Uniform selection of subsets with replacement*:

$$C_T = [1 - (1 - 1/N)^{n1}] \cdot \theta^{n3}$$

b) *Uniform selection without replacement*:

$$C_T = (n2/N) \cdot \theta^{n3}.$$

AN EXAMPLE COMPARING UNIFORM AND OPERATIONAL FAILURE RATES

An interesting example based on real applications may help to realize the impact of a uniform test on operational reliability. It is based on data concerning 9 large IBM software products, published by Currit, Dyer and Mills (1986).

In spite of the product differences, the informations on their failure rates (shown in Table 1) are astonishingly consistent. They relate to 8 classes of mean times to failure (ranging from 19 months to 5 000 years) the percentage of related errors found.

Table 1 Mean time to failure occurrence
Currit, Dyer and Mills (1986)

MEAN TIME TO PROBLEM OCCURRENCE IN KMONTHS BY RATE CLASS

PRODUCT	60	19	6	1.9	.6	.19	.06	.019
1	34.2	28.8	17.8	10.3	5.0	2.1	1.2	0.7
2	34.2	28.0	18.2	9.7	4.5	3.2	1.5	0.7
3	33.7	28.5	18.0	8.7	6.5	2.8	1.4	0.4
4	34.2	28.5	18.7	11.9	4.4	2.0	0.3	0.1
5	34.2	28.5	18.4	9.4	4.4	2.9	1.4	0.7
6	32.0	28.2	20.1	11.5	5.0	2.1	0.8	0.3
7	34.0	28.5	18.5	9.9	4.5	2.7	1.4	0.6
8	31.9	27.1	18.4	11.1	6.5	2.7	1.4	1.1
9	31.2	27.6	20.4	12.8	5.6	1.9	0.5	0.0

In particular we note that only 2 % of all errors (both highest classes) caused a thousand times more failures per error than the two lowest classes, which account for about 60 % of all of them.

In order to simplify our analysis, we coarsen the granularity of Table 2 by merging any 2 adjacent columns, thus obtaining the following approximative values.

For an average failure rate x corresponding to a MTTF between 19 and 60 months the experimental data show that:

about 60 % of all errors contribute to an average rate $x \cdot 10^{-3}$;

about 30 % of all errors contribute to an average rate $x \cdot 10^{-2}$;

about 8 % of all errors contribute to an average rate $x \cdot 10^{-1}$;

about 2 % of all errors contribute to an average rate x.

Denoting by y the (unknown) number of errors and assuming disjoint failure sets, this means that the momentary failure rate at time t_0 is:

$$\lambda(t_0) \cong (0.6 \cdot 10^{-3} + 0.3 \cdot 10^{-2} + 0.08 \cdot 10^{-1} + 0.02) \cdot x \cdot y$$

$$= 0.0316 \cdot x \cdot y$$

A uniform test T with constant detectability C_T will only reduce the absolute number of errors, leaving their proportionality invariant:

$$E[\ \lambda(\text{after test T})\ ] \cong 0.0316 \cdot x \cdot y \cdot (1 - C_T)$$

A statistical testing simulating operation, on the other hand, will detect errors according to their rate classification, i.e. after the first error correction the failure rate will have decreased by:

$$E[\Delta\lambda] := \lambda(t_0) - E[\lambda(\text{after first operational failure time} > t_0] \cong$$

$$\cong (0.02 \cdot 10^3 + 0.08 \cdot 10 + 0.3 \cdot 10^{-1} + 0.6 \cdot 10^{-3}) \cdot x / 31.6 \cong$$

$$\cong 0.659 \cdot x$$

This means that at present the improvement promised by statistical testing is considerably higher than the uniform test would let expect (to achieve the same order of magnitude, the uniform test should detect about 20 errors).

The main draw-back of simulation, however, lies in the temporal development of $\Delta\lambda$: as failures are detected, the number of high-rate errors will decrease, thus reducing the expected improvement given by $\Delta\lambda$. Let us assume that exactly all (and only) the 2 % errors of the high rate category have been removed at time t_1. The failure rate is then:

$$\lambda(t_1) \cong (8 \cdot 10^{-3} + 3 \cdot 10^{-3} + 0.6 \cdot 10^{-3}) \cdot x \cdot y \cong 0.0116 \cdot x \cdot y$$

The improvement expected to be achieved by a failure detection by statistical testing at this time is:

$$E[\Delta\lambda'] := \lambda(t_1) - E[\lambda(\text{after first operational failure time} > t_1)] \cong$$

$$\cong (8 \cdot 10^{-1} + 3 \cdot 10^{-2} + 0.6 \cdot 10^{-3}) \cdot x / 11.6 \cong 0.0716 \cdot x$$

This shows the decreasing efficiency of statistical testing over time.

SYSTEMATIC TEST DIVERSITY

The process described in the previous sections may be obviously extended allowing more than one systematic test to be integrated into the overall statistical testing phase.

By definition, the successive performance of two uniform tests T_1 and T_2 with error detection probabilities C_1 resp. C_2 will result again in an uniform test $T := T_1 \cup T_2$ with higher probability C_T of detecting an arbitrary error.

The proportionality shown above

$$\lambda_T(t) = \lambda_0(t) \cdot U_T,$$

with $U_T = 1 - C_T$,

will then analogously hold for the single tests T_1 resp. T_2:

$$\lambda_i(t) = \lambda_0(t) \cdot U_i,$$

with $U_i = 1 - C_i$, $i \in \{1,2\}$.

As such a test diversity is successful as soon as errors are detected by one of both alternatives T_1 or T_2, it can be easily compared with a 1-out-of-2 system failing if and only if both versions fail:

$$U_T = 1 - C_1 - C_2 + C_{12},$$

where C_{12} is the probability of an error being detected by both tests.

In particular, similarly to version dependence, we may also expect in case of diverse tests a dependent detection behaviour, which in case of disjoint sets of corrected failures may even degenerate in a situation "better than independence".

Anyway, the correlation between two testing phases may be represented by a factor α_{12} with

$$U_T = U_1 \cdot U_2 \cdot \alpha_{12}.$$

The dependence factor α_{12} might be possibly evaluated similarly as for version dependence (s. Saglietti (1990)) by means of a comparison of structural coverage measures achieved by the diverse techniques with respect to the same input selection. The deviation of the resulting coverage growth curves, automatically supplied by a number of supporting tools, provide information on the (dis-) similarity in the testing behaviour of the alternative strategies.

Another aspect which can be easily translated from the theory of version diversity into a theory of "testing diversity" regards the improvement expected to be achieved by enforcing dissimilarity during the process of selecting test cases.

Here we could interpret "ad hoc" - diversity as testing phases possibly performed with respect to identical or similar strategy: different testing teams acting independently may just increase the time of the same testing method or replicate simulations of the same random distribution.

On the other hand, forced testing diversity may be achieved by choosing test cases according to different selection methods given by diverse strategies, as static or dynamic checks, functional or structural distributions, inspections based on the control flow or on the data flow.

The well-known results of Littlewood and Miller (1987) may be now extended on systematic strategies expected to be equally effective. They provide formal support to the intuitive expectation of increasing efficiency by forcing design dissimilarity.

Experimental evaluations of testing diversity have indirectly already been performed in a number of studies, for example by Basili and Selby (1987) as well as Ntafos (1988).

A more recent experiment was presented by Shimeall and Leveson (1989). In the context of this contribution, the particular value of their work lies in reporting the number of faults having been commonly detected by different strategies and of those having been discovered by only a particular one of them.

Without going into the details of this investigation, which can be read in the referenced publication, we recall in Table 2 the main results reported, which already provide a real-world example for the testing dependence factors introduced above.

Table 2 Number of faults detected Shimeall and Leveson (1989).

Method	Version 1	2	3	4	5	6	7	8	Total
Testing only	2	12	21	11	13	1	11	10	81
2-Version voting only	10/11	9/9	11/12	7/8	14/14	7/7	8/8	10/10	72/78
Code reading only	0	2	4	2	0	1	0	16	25
Assertions only	3	3	1	8	1	1	3	3	23
Static analysis only	0	0	2	0	0	0	0	0	2
Both 2-v. voting & test	3	1	1	3	2	6	4	0	20
Both assertions & test	5	1	0	2	5	3	0	0	16
Both assertions & 2-v. voting	0	0	2	0	2	0	4	4	12
Both reading & assertions	2	0	0	1	1	0	0	0	4
Both static analysis & 2-v. voting	0	0	0	0	0	1	1	0	2
Both reading & 2-v. voting	0	0	2	0	0	0	0	0	2
Both reading & test	0	0	0	0	0	0	1	0	1
Both static analysis & test	0	0	0	0	0	0	1	0	1
Assert & 2-v. voting & test	0	0	1	1	2	1	0	0	5
Reading & 2-v. voting & test	0	2	0	0	0	0	0	2	4

In this table the quality of a specific checking method is not just evaluated in terms of the absolute number of errors found or features covered, but it essentially results from the intrinsic diversity with respect to the alternative techniques.

Should such data be collected in significant quantities in future and should they provide evidence for problem-independent consistencies, this certainly would increase the feasibility of the approach proposed above for the combination of testing phases.

BEST CHOICE AMONG SYSTEMATIC TESTS

In spite of the improvements generally expected from diversified tests, the effort required may possibly be too high, so that one has perhaps to choose among two different checking strategies T_1 and T_2.

Let us assume that we have already estimated the operational hazard rate $\lambda(t)$, which is still much higher than the required level λ_{max}. On the basis of past failures and coverage curves we may be able to obtain rough estimates of the proportionality constants U_1 resp. U_2. Then we have the following alternatives:

1) *Strategy S_1*: combine a statistical test of duration t_1 such that:

$$\lambda(t_1) = U_1^{-1} \cdot \lambda_{max}$$

with a systematic test T_1;

or, alternatively:

2) *Strategy S_2*: combine a statistical test of duration t_2 such that:

$$\lambda(t_2) = U_2^{-1} \cdot \lambda_{max}$$

with a systematic test T_2.

The logarithmic Poisson execution time model for software reliability measurement proposed by Musa and Okumoto (1984) is based on the random process representing the number $M(t)$ of failures experienced by execution time t in program P.

The mean value:

$$\mu(t) := E[M(t)]$$

is then used to define the *failure intensity function*:

$$\lambda(t) := d\mu(t) / dt,$$

which by assumption decreases exponentially with the expected number of failures experienced:

$$\lambda(t) = \lambda_o \cdot \exp(-\theta \cdot \mu(t)),$$

where λ_o and θ are model parameters to be estimated on the basis of failure history.

By substitution (s. Musa and Okumoto (1984)) we obtain:

$$\mu(t) = (\ln(\lambda_o \cdot \theta \cdot t + 1)) / \theta$$

$$\Rightarrow \lambda(t) = \lambda_o / (\lambda_o \cdot \theta \cdot t + 1)$$

The expected time t_1 respectively t_2 necessary to achieve the desired reliability level in each case can be therefore easily determined by:

$$\lambda_o / (\lambda_o \cdot \theta \cdot t_i + 1) = U_i^{-1} \cdot \lambda_{max}$$

$$\Rightarrow t_i = (\lambda_o \cdot U_i / \lambda_{max} - 1) / (\lambda_o \cdot \theta),$$

$$\text{with } i \in \{1,2\}.$$

Assuming a cost K for each further time unity of statistical test as well as a total overhead K_1 respectively K_2 for each systematic phase, we obtain the following alternative efforts $K(S_1)$ and $K(S_2)$ for both strategies proposed above:

$$K(S_i) = K \cdot t_i + K_i =$$

$$= K \cdot (\lambda_o \cdot U_i / \lambda_{max} - 1) / (\lambda_o \cdot \theta) + K_i,$$

$$\text{with } i \in \{1,2\}$$

The optimal choice among uniform testing strategies is then implied by comparing the expected costs:

$$K(S_1) > K(S_2)$$

$$\Leftrightarrow U_1 - U_2 > (K_2 - K_1) \cdot \lambda_{max} \cdot \theta / K.$$

CONCLUSION

A particular class of systematic testing strategies was shown to be suitably treatable as explanatory variable of Proportional Hazard Models: they establish the relationship between the base-line failure rate before testing and the resulting failure rate obtained after a checking phase, during which each error was removed with equal probability.

All considerations presented are based on average observations: they regard the single error detection probabilities as being comparable, if the sequence of the related failure occurrences is not strictly dependent on time. In fact, by selecting the test cases mainly according to deterministic input partitions, we can reduce the importance of "failure size", until no trend in time is observable, thus averaging the chance of each error being detected.

The main difficulty of this technique obviously consists of estimating the constant discovery probability per error C_T. Some possible approximations were suggested.

The main concepts presented were analyzed within the context of real-world data. Among the possible applications of this theory both the combination and the optimal choice of different systematic testing criteria were considered.

REFERENCES

Ascher, H. (1986). The Use of Regression Techniques for Matching Reliability Models to the Real World. *NATO ASI Series, Vol. F22.*

Basili, V.R. and Selby, R.W. (1987). Comparing the Effectiveness of Software Testing Strategies. *IEEE Transactions on Software Engineering, Vol SE-13,No.12.*

Currit, P.A., Dyer, M. and Mills, H.D. (1986). Certifying the Reliability of Software. *IEEE Transactions on Software Engineering, Vol. SE-12, No.1.*

Dale, C.J. (1985). Application of the Proportional Hazards Model in the Reliability Field. *Reliability Engineering 10 (1985), 1 - 14.*

Kalbfleisch, J.D. and Prentice, R.L. (1980). The Statistical Analysis of Failure Time Data (chapter 4: The Proportional Hazards Model). *Wiley Series in Probability and Mathematical Statistics*

Feller, W. (1957). An Introduction to Probability Theory and its Application. *Vol. 1, Wiley, New York.*

Littlewood, B. (1987/1988). Private communications at ESPRIT reviews.

Littlewood, B. and Miller, D.R. (1987). A Conceptual Model of Multi-Version Software. *17th Int. Symp. on Fault-Tolerant Computing (FTCS-17), IEEE Computer Society Press.*

Mills, H.D. (1970). On the Statistical Validation of Computer Programs. *IBM FSD Rep.*

Mourad, S. and Andrews, D. (1987). On the Reliability of the IBM MVS/XA Operating System. *IEEE Transactions on Software Engineering, Vol. SE-13, No.10.*

Musa, J.D. and Okumoto, K. (1984). A. Logarithmic Poisson Execution Time Model for Software Reliability Measurement. *IEEE Computer Society Press.*

Ntafos, S.C. (1988). A Comparison of Some Structural Testing Strategies. *IEEE Transactions on Software Engineering, Vol. 14, No. 6.*

Puhr-Westerheide, P. and Krzykacz, B. (1982). A Statistical Method for the Detection of Software Errors. *IFAC Conf. on Software for Computer Control,Madrid, Spain.*

Rossetti, D.J. and Iyer, R.K. (1982). Software Related Failures on the IBM 3081: A Relationship with System Utilization. *Center for Reliable Computing, Stanford University, CRC Technical Report No. 82-8.*

Saglietti, F. (1988). Optimal Combination of Software Testing Strategies. *IFAC/ IFIP Int. Symp. on Safety Related Computers (SAFECOMP '88), Fulda, Germany.*

Saglietti, F. (1990). Measurement of Diversity Degree by Quantification of Dissimilarity in the Input Partition. *IEE Software Engineering Journal, Vol. 5, No. 1.*

Shimeall, T.J. and Leveson, N.G. (1989). An Empirical Comparison of Software Fault Tolerance and Fault Elimination. *Naval Postgraduate School, Monterey California, Rep. NPS 52-89-047.*

Wightman, D.W. and Bendell, A. (1986). Proportional Hazards Modelling of Software Failure Data. *Trent Polytechnic Nottingham, UK.*

THE ROLE OF TREND ANALYSIS IN SOFTWARE DEVELOPMENT AND VALIDATION[*]

K. Kanoun and J.-C. Laprie

LAAS-CNRS, 7 Avenue du Colonel Roche, 31077 Toulouse, FRANCE

Abstract: The aim of this paper is to show how reliability trend analysis can help the designer in controlling the progress of the development activities and appreciating the efficiency of the test programs. Software reliability growth is first characterized and practical recommendations for trend analysis are discussed. Application of trend tests to some data sets collected on real systems illustrates the proposed method.

Introduction

Generally Software reliability studies are based on reliability growth models application in order to evaluate the reliability measures. When performed for a large base of deployed software systems, the results are usually of high relevance (see e.g. [Ada 84, Kan 87] for examples of such studies). However, utilization of reliability growth models during early stages of development and validation is much less convincing: when the observed times to failure are of the order of magnitude of minutes or hours, the predictions performed from such data can hardly predict mean times to failure different from minutes or hours ... which is so distant of any expected reasonable reliability as is not very helpful to perform such estimations. In addition, when a program under validation becomes reliable enough, the times to failure may simply be large enough in order to make the application of reliability growth models impractical, due to the (hoped for) scarcity of failure data. On the other hand, in order to become a true engineering exercise, software validation should be guided by quantified considerations relating to its reliability. Statistical trend tests provide for such guides.

This paper is devoted to the presentation of trend tests which are intended to help the management of the development and validation process. It will be shown that, for several circumstances, trend tests give information of prime importance to the developer. Emphasis will be put on the way they can be used during project progress and on practical results that can be derived from their use.

The paper is composed of three sections. The first section is devoted to the characterization of reliability growth. In the second section, trend tests are presented and discussed; the type of results which can be drawn from trend analysis are stated. The third section is devoted to exemplifying the results from the first and second section on failure data collected on real systems.

1. Reliability growth characterization

1.1. Practical considerations

Software lack of reliability stems from the presence of faults, and is manifested by failures which are consecutive to fault sensitization[1]. Removing faults should result in reliability growth. However, it is not always so, due to the complexity of the relation between faults and failures, thus between faults and reliability, which has been noticed a long time ago (see e.g. [Lit 79]). Basically, complexity arises from a double uncertainty: the presence of faults, the fault sensitization via the trajectory in the input space of a program[2]. As a consequence, one usually observes reliability trend changes, which may result from a great variety of phenomena, such as:

- variation in the utilization environment: the variation in the testing effort during debugging, change in test sets, addition of new users during the operational life, etc.,

---

[*] This work was supported in part by the ESPRIT BRA project 3092 "Predictably Dependable Computing System" (PDCS).

---

[1] Precise definition of faults, failures, reliability, etc. are given in [Lap 87].

[2] As an example, data published in [Ada 84] concerning nine large software products show that for a program with a mean lifetime of fifteen years, only 5% of the faults will be activated during this period.

- dependency of faults: some software faults can be masked by others, i.e. they cannot be activated as long as the latter are not removed [Ohb 84],
- variation in time delay between the detection of an error and its removal; this is closely dependent on the nature of the activated faults: some faults are more difficult to identify than others and take longer time to be removed.

With this in mind, reliability decrease may not, and usually does not, mean that the software has more and more faults; it does just tell that the software exercises more and more failures per unit of time under the corresponding conditions of use. Corrections may reduce the failure input domain but more faults are activated or faults are activated more frequently. However, during fault correction new faults may be also introduced — regression faults — which can deteriorate or not software reliability depending on the conditions of use. Last but not least, reliability decrease may be consecutive to specification changes.

1.2. Formal definitions

From what precedes, it can be seen that software reliability may be characterized by means of two types of random variables: the inter-failure time or the number of failures per unit of time (i.e. the failure intensity). These two random variables are not independent: knowing the inter-failure times it is possible to obtain the failure intensity (the second form needs less precise data collection). Both of them are considered in our work. The choice between one variable or the other may be guided by the following elements: i) the objective of the reliability study (development follow up, maintenance planning or reliability evaluation), ii) the way data is collected and iii) the life cycle phase concerned by the study.

Let T_1, T_2, ... denote the sequence of random variables corresponding to inter-failure times, and $F_{T_i}(x)$ the distribution function of T_i. Reliability growth is characterized by the fact that inter-failure times tend to become larger, i.e.:

$$T_j \underset{st}{\leq} T_i, \quad \text{for all } j < i \qquad (1)$$

where $\underset{st}{\leq}$ means stochastically smaller than; under the stochastic independency assumption this is equivalent to:

$$F_{T_j}(x) \geq F_{T_i}(x) \quad \text{for all } j < i \text{ and } x \qquad (2)$$

Data collection in the form of inter-failure times may be tedious mainly during development, in which case it is more suitable and less time consuming to collect data in the form of number of failures per unit of time. The unit of time is function of the type of use of the system as well as the number of failures occurring during the considered units of time.

In this case, reliability growth is expressed by:

$$N(t_1)+N(t_2) \underset{st}{\geq} N(t_1+t_2) \quad \text{for all } t_1, t_2 \geq 0 \qquad (3)$$

where $N(t)$ is the number of observed failures during time interval $[0,t]$. Inequality (3) must be strict for at least a couple (t_1,t_2) and means that the expected number of failures in any initial interval $[0,t_2]$ is no smaller than the expected number of failures in any interval of the same length occurring later $[t_1,t_1+t_2]$. Let $H(t)$ denote its expectation: $E[N(t)]$. Assuming that the failure process has independent increments leads to:

$$H(t_1)+H(t_2) \geq H(t_1+t_2) \quad \text{for all } t_1, t_2 \geq 0 \qquad (4)$$

When (4) holds, $H(t)$ is said to be a subadditive function [Hol 74]; when it is reversed $H(t)$ is said to be superadditive and denotes reliability decrease.

2. Reliability growth analysis

Reliability growth can be analysed through the use of trend tests: these tests give a better insights into the evolution of the reliability. Several trend tests may be employed for each type of random variable; due to space limitation, only the most used and significant ones are presented in this section[3]. The presentation of the tests is followed by a discussion on how they can be used for studying software reliability. Some types of results that can be drawn from trend analysis are discussed in the last sub-section.

2.1. Trend tests presentation

2.1.1. Inter-failure times

Among the existing tests, the arithmetical mean and the Laplace tests can be used. The first test consists of calculating τ_k the arithmetical mean of the first k observed inter-failure times θ_i (which are the realizations of T_i, i = 1, 2, ..., k):

$$\tau_k = \frac{1}{k} \sum_{i=1}^{k} \theta_i \qquad (5)$$

When τ_k form an increasing series, reliability growth is deduced. This test is very simple and is directly related to the observed data. It is a graphical test and as such is informal.

A more rigourous test consists of calculating the Laplace factor [Cox 66] for the observation period t_o, $u(t_o)$. The occurrence of the events is assumed to follow a non-homogeneous Poisson process (NHPP) whose failure intensity is decreasing and is given by:

$$h(t) = e^{a+bt} \qquad b < 0 \qquad (6)$$

If b=0 the Poisson process becomes homogeneous and the occurrence rate is time independent.

---

3 For a more complete presentation and discussion on trend tests see e.g. [Asc 84, Gau 90].

Under this hypothesis (b=0), the statistics:

$$u(t_o) = \frac{\dfrac{1}{n}\displaystyle\sum_{i=1}^{n} s_i - \dfrac{t_o}{2}}{t_o\sqrt{\dfrac{1}{12\,n}}} \qquad (7)$$

(where n is the number of failures in $[0,t_o]$, and s_i the time of occurrence of failure i, i=1,.., n)

is approximately normal distributed with zero mean and unit variance. Negative value of $u(t_o)$ means that the considered statistics is below the mean and therefore indicates b<0, i.e. a decreasing failure intensity. On the other hand, positive values suggest an increasing failure intensity.

2.1.2. Failure intensity

Two very simple graphical tests can be used (the plot giving the evolution of the observed cumulative number of failures or failure intensity versus time) as well as some analytical tests (among which the Laplace test and the superadditive test). These four tests are briefly presented and discussed hereafter.

Graphical tests

Figure 1 gives the observed cumulative number of failures and the corresponding trend: this trend is directly related to relation (3); a concave curve (down) indicates reliability growth (i.e. inequality (3) holds) and, in the converse, a convex curve results in reliability decrease. The observed number of failures per unit of time n_k (the failure intensity) corresponding to these situations is given in figure 2 ($N_k = \displaystyle\sum_{i=1}^{k} n_i$).

The Laplace test

Following the method outlined in [Cox 66], the Laplace factor is derived in [Kan 91] and is given by:

$$u(k) = \frac{\dfrac{1}{N_k}\displaystyle\sum_{i=1}^{k} (i-1)n_i - \dfrac{(k-1)}{2}}{\sqrt{\dfrac{(k^2-1)}{12\,N_k}}} \qquad (8)$$

As previously, negative values of u(k) indicate reliability growth whereas positive values indicate reliability decrease.

The superadditive test

The Laplace test is well suited to test monotonic trend versus no trend; when the collected data do not exhibit monotonic trend, the superadditive test is more adapted as discussed in [Hol 74, Ash 84]. This test is based on relations (3) and (4): when these relations hold reliability growth is deduced.

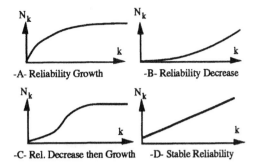

Figure 1: Cumulative number of failures and reliability evolution

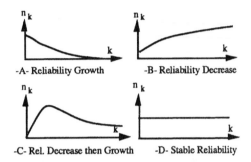

Figure 2: Failure intensity and reliability evolution

2.2. Discussion, practical recommendations

Comparison of several trend tests among which the Laplace test and the superadditive test according to optimality and/or consistency is carried out in [Hol 74, Ash 84, Gau 90]. The latter reference recommends the use of the Laplace tests in several situations mainly when processing raw data or under the NHPP assumption. However it is not well adapted to identify reliability fluctuation and even trend changes such as situation c of figures 1 and 2. Indeed it is well suited to test monotonic trend only. On the other hand, the superadditive test is more adapted to non-monotonic trend, but it is more difficult to be used in a systematic way. The Laplace test suffers from the fact that it is specific of a given model (expression (6)) and one has to associate a specific expression of this factor to each model.

For our purpose, we adopt a pragmatic point of view: we do not use the Laplace test as a statistics with confidence interval, but merely as an indicator to test the trend. Moreover the investigation of the evolution of this factor will help in detecting trend fluctuation. Actually, this factor can be evaluated step by step (at each unit of time or after each failure) and trend change of this factor indicates *local* trend change in the data. This is illustrated in figure 3-a: Considering for instance only data from the trend change point A leads to negative values (i.e. reliability growth) as indicated in figure 3-b. Periods of reliability growth and decrease can thus be identified.

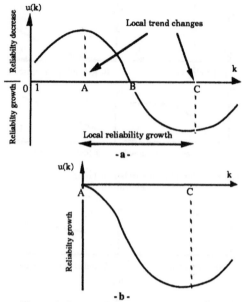

Figure 3: Laplace factor and local fluctuation

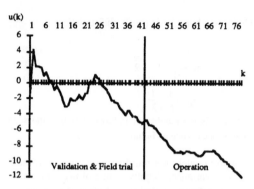

Figure 4: Laplace factor for the TROPICO-R considering the whole data set

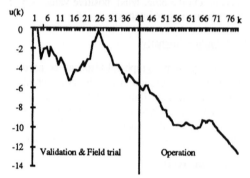

Figure 5: Laplace factor for TROPICO-R without considering the first three failure data

This phenomena is illustrated by figures 4 and 5 which relate to the TROPICO-R switching system studied in [Kan 91]. Figure 4 gives the Laplace factor for the whole data set from validation to operation. At the beginning of the validation, reliability decrease took place, as a result of the correction of 28 faults during the third unit of time whereas only 8 faults were removed during the first two time units and 24 during the next two time units; applying the trend test without the data belonging to the three first units of time leads to reliability growth (figure 5).

In real situations, we will use the Laplace test to analyse the trend considering the sign of its factor as well as the evolution of this factor with time. A Laplace factor oscillating around a constant value (within a bound of -2 and +2 implies) stable reliability.

2.3. Results which can be issued from trend analysis

Trend analyses are of great help in appreciating the efficiency of test activities and controlling their progress. They help considerably the software development follow up. Indeed graphical tests are more often used in the industrial field [Gra 87, Lev 91, Val 88].

Reliability decrease at the beginning of a new activity such as i) new life cycle phase, ii) change in the test sets within the same phase, iii) adding of new users or iv) activating the system in a different profile of use, etc., is generally expected and is considered as a normal situation. Reliability decrease may also result from regression faults. Trend tests allow to detect this kind of behavior. If the duration of the period of decrease seems long, one has to pay attention and, in some situations, if it keeps decreasing

this can point out some problems within the software: the analysis of the reasons of this decrease as well as the nature of the activated faults is of prime importance in such situations. Such analysis may help in the decision to re-examine the corresponding piece of software.

Reliability growth after reliability decrease is usually welcomed since it indicates that, after first faults removal, the corresponding activity reveals less and less faults. When calendar time is used, mainly in operational life, sudden reliability growth may result from a period of time during which the system is less used or is not used at all; it may also result from the fact that some failures are not recorded. When such situation is noticed, one has to be very careful and, more important, an examination of the reasons of this sudden increase is essential.

Stable reliability indicates that the corresponding activity has reached a "saturation": application of the corresponding tests set does not reveal new faults, or the corrective actions performed are of no perceptible effect on reliability; one has either to stop testing or to introduce new sets of tests or to proceed to the next phase. More generally a test set has to continue to be applied as long as it exhibits reliability growth and stopped when stable reliability is reached.

Finally, it is noteworthy that trend analyses may be of great help for reliability growth models to give better predictions since they can be applied to data displaying trend in accordance with their assumptions: failure data can be partitioned according to the trend and two types of reliability growth models can be applied: i) when the data exhibit reliability decrease followed by reliability growth, an S-Shaped model [Ohb 84] can be applied, ii) in case of reliability growth most of the other existing reliability growth models can be applied.

3. Application to real systems

Four different systems are considered:
- the first one, called **system A**, corresponds to a system which has been observed during validation and a part of operational life [Met 90],
- the second, to system 27 published in [Mus 79], called **system B** hereafter,
- the third one is also issued from [Mus 79], system SS4, called **system C**,
- the last one corresponds to the system considered in section 2 [Kan 91] and is called **system D**.

3.1. System A

The Laplace factor for this system is given in figure 6. System A displayed reliability decrease during the validation, reliability growth took place during operational life only. This is confirmed by figure 7 where the Laplace test is applied separately to each phase. It can also be seen that some reliability fluctuations took place from unit time 15, this fluctuation is due to introduction of new users.

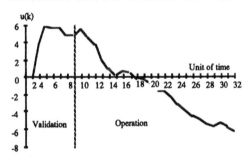

Figure 6: Laplace factor for System A considering the whole data set

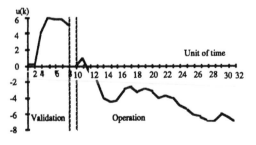

Figure 7: Laplace factor for System A considering each phase separately

Clearly, applying reliability growth models during validation would not have given helpful insights. An S-Shaped model can be applied to the whole data set and any reliability growth model to operational data.

3.2. System B

System B is an example of systems which exhibit two phases of stable reliability; transition between them took place about failures 23-24 (figure 8). This system was under test and one has to know what happened at this time and why no reliability growth took place. It was not possible from the published data to identify the reasons of this behavior. In this case, data may be partitioned into two subsets each of them being modeled by a constant failure rate: the failure rate of the second subset (from 24 to 42 being lower than the failure rate of the first subset).

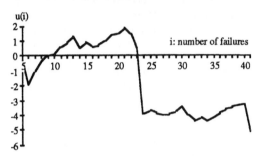

Figure 8: Laplace factor for System B

3.3. System C

Data gathered on System C correspond to operational life. Application of the arithmetical mean in figure 9 shows that the mean time to failure is almost constant: it is about $230 \ 10^3$. The corresponding Laplace factor oscillates between -2 and +2 indicating also stable reliability. In this case, a constant failure rate is well adapted to model the software behavior and is of simpler application than a reliability growth model. This result is not surprising since the software was not maintained (no fault correction).

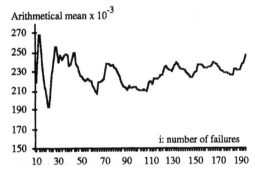

Figure 9: Arithmetical mean for System C

3.4. System D

Trend tests for this system are displayed in figures 4 and 5. Trend tests applied separately to each phase is illustrated in figure 10.

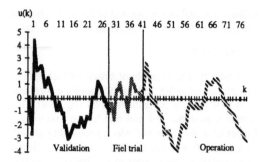

Figure 10 - Laplace factor of System D for each phase

It is interesting to comment figures 4 and 10 at the same time:
- the local reliability decrease from k=14 to k=25, was induced by the changes in nature of the tests within the validation phase: this period corresponds to the application of *quality* and *performance* tests after *functional* tests in the previous period; this decrease is due to their dynamic nature (traffic simulation) which activated new parts of the program,
- transitions from validation to field trial and from field trial to operation did not give rise to a reliability discontinuity, which means that the tests applied during the end of validation are representative of operational conditions,
- figure 4 indicates that from k=55 up to k=70 reliability tends to be stabilized; u(k) is almost constant, suggesting a local reliability decrease; this behavior is reinforced when considering the trend results obtained for operational data only in figure 10 where reliability decrease is more evident; from k=70 this trend is reversed; this failure behavior is directly related to the number of installed exchanges over the periods considered, during which about twelve exchanges were installed and the number of failures reported by the users increased; by time unit 70, a new system had been released and no additional former system had been installed, which corresponds to the period of local reliability growth.

Applying reliability growth models blindly to this data set would have conducted to non significant results, on the other hand, using trend analysis results leads to trustworthy results [Kan 91].

Conclusion

Trend analyses constitute a major tool during the software development process. It has been shown how the results can guide the designer to control the progress of the development activities and even to take the decision to re-examine the software. Trend analyses are also of prime importance when reliability evaluation is needed. They allow periods of times exhibiting reliability growth and reliability decrease to be identified in order to apply reliability

growth models to data exhibiting trend in accordance with their modeling assumptions. Trend tests and reliability growth models are part of a global method for software reliability analysis and evaluation which is presented in [Kan 88] and has been applied successfully to data collected on real systems [Kan 87, Kan 91].

References

Ada 84 E.N. Adams, "Optimizing preventive service of software products", *IBM J. of Research and Development*, vol. 28, no. 1, Jan. 1984, pp. 2-14.

Asc 84 H.Ascher, H.Feingold, *Repairable Systems Reliability: Modeling, Inference, Misconceptions and Their Causes*, Lecture notes in statistics, Vol. 7, 1984.

Cox 66 D.R.Cox, P.A.W.Lewis, *The Statistical Analysis of Series of Events*, London, Chapman & Hall, 1966.

Hol 74 M.Hollander, F.Prochan, " A test for superadditivity for the mean value function of a Non Homogeneous Poisson Process", *Stoch. Proc. and their application* , vol. 2, 1974, pp. 195-209.

Gau 90 O.Gaudoin, "Statistical tools for software reliability evaluation", Phd thesis, Joseph Fournier Univ. Grenoble I, Dec. 1990, in French.

Gra 87 R.B.Grady, D.R.Caswell, *Software metrics: establishing a company-wide program*, Hewlett Packard Company, Prentice Hall, Inc 1987.

Kan 87 K.Kanoun, T.Sabourin, "Software Dependability of a Telephone Switching System", Proc. 17th *IEEE Int. Symp. on Fault-Tolerant Comp. (FTCS-17)*, Pittsburgh, Pennsylvania, July, 1987, pp. 236-241.

Kan 88 K.Kanoun, J.C.Laprie, T.Sabourin, "A Method for Software Reliability Growth Analysis and Assessment", Proc. of *Software engineering &its applications*, Toulouse, France, Dec. 1988, pp. 859-878.

Kan 91 K.Kanoun, M.Bastos Martini, J.Moreira De Souza, "A Method for Software reliability analysis and Prediction — Application to The TROPICO-R Switching System", *IEEE Transactions on Software Engineering*, April 1991, pp. 334-344.

Lap 87 J.C.Laprie: "Dependability: a unifying concept for reliable computing and fault tolerance", *Resilient Computing Systems*, Vol. 2, T.Anderson Editor, Collins et Wiley, 1987

Lev 91 Y.Levendel, "Software quality improvement process: when to stop testing", Proc. of *Software engineering and its applications*, Toulouse, France, Dec. 1991, pp. 729-749.

Lit 79 B.Littlewood, "How to measure software reliability and how not to", *IEEE Tranactions on Reliability*, vol. R-28, no. 2, June 1979, pp. 103-110.

Met 90 S.Metge, "Reliability analysis and evaluation of two telecommunications software", LAAS research report n° 90.112, May 1990, in French.

Mus 79 J.D.Musa, "Software Reliability data", Data and Analysis Centre for Software Rome Air Development Centre (RADC) Rome, NY, 1979.

Ohb 84 M.Ohba, S.Yamada, "S-Shaped Software Reliability Growth Models", Proc. *4 th International Conference on Reliability and Maintainability*, Perros Guirec, France, 1984, pp. 430-436.

Val 88 V.Valette,"An environment for software reliability evaluation", Proc. of *Software engineering & its applications*, Toulouse, France, Dec. 1988, pp. 879-897

THE BALANCING OF QUALITY ASSURANCE AGAINST VALIDATION ACTIVITIES

T. Stålhane

SINTEF DELAB, 7034 Trondheim, Norway

Abstract. This paper discusses the application of a quality assurance effect model to:
- assign test-equivalent values to quality assurance activities
- compare alternative sets of quality assurance activities with respect to their effect on reliability and project costs.

The results indicate that fault avoidance techniques are much more effective than fault detection techniques. By using Bayes method, it is possible to use the effect of the quality assurance activities to find a prior distribution for the systems reliability and thus achieve the required confidence with a smaller acceptance test suite than is otherwise needed.

Keywords. Software Engineering, Reliability, Quality Control, Bayes Methods.

INTRODUCTION

Before a piece of software is delivered to the customer it is necessary to run a validation test. The purpose of this test is to give the customer confidence that the system meets all requirement, among them reliability.

In most cases, both for software and for hardware, confidence in the quality of the product stems from many sources. Among these, we will consider two; the amount of quality assurance and the amount of testing.

The goal of this paper is to show a way to combine the results of tests and quality assurance activities in order to achieve the right degree of confidence and how it is possible to combine the two activities to achieve an optimum balance.

THE PROBLEM

As all other quality attributes, reliability is built in during production. This paper does not, however, deal with the reliability per se, but with our confidence that the product has the promised reliability. Simply put, our question is as follows:

Given the amount of quality assurance done during production, how many tests do we need to run in order to feel sure that we have a product with the required reliability?

To resort to the language of the disarmement negotiatons: "Trust, but verify", or in our case, "Trust but validate".

A populare approach has been to get confidence by running a validation test suite.

Often, the results of this validation is interpreted alone. That is: We assume absolute ignorance before we run any tests and the assessment of the fulfillment of the requirements is done on the basis of the test results alone.

The question is now, given that we run X tests and none of them fail, what is our assessment of the systems reliability? Or more directly, how confident are we that the system will function without errors during the next, say 1000 executions?

Under the assumption of independent inputs that are unformly distributed over the input space and a constant failure rate, we can find the number of successful executions needed to give us a say 95% confidence that the reliability is better than 0.999. The problem is, however, that this gives us the impractically large number of 3000 tests.

The following table is taken from Parnas (1990):

N	$M=(1 - 1/h)^N$
500	0.60638
600	0.54865
700	0.49641
800	0.44915
900	0.40639
1000	0.36700
1500	0.22296
2000	0.13520
2500	0.08198
3000	0.04971
3500	0.03014
4000	0.01828
4500	0.01108
4700	0.00907
5000	0.00672

$h=1000$

Table 1. Probablility that a system with failure probability of .001 will pass N successive tests

In order to get a lower number of necessary test cases, we need to take into account the amount of quality assurance done during production. It seems reasonable to assume that a customer will have greater confidence in a system with no errors in 100 test cases and a large amount of quality assurance, than in a system with the same test result, but put together in an unstructured and haphazard way.

A practical way to include prior knowledge is to use a Bayesian model. In this way, we can include our experience with the producer and our knowledge of the amount of quality assurance done by building a prior distribution. This will be discussed in the next chapter.

In the rest of this paper, we will use the following notation:

R : the systems reliability

X : the number of tests run

N : the number of failures observed during the N tests.

THE BASIC MODEL

It is reasonable, at least as an approximation, to use the following assumptions:

$N \sim Bin (X, 1-R)$

$R \sim Beta(\alpha, \beta)$

From this, it follows that

$$R|N,X \sim Beta (\alpha+X-N, \beta+N) \qquad (1)$$

Our posterior estimate for R is then

$$R^* = \frac{\alpha + X - N}{\alpha + \beta + X} \qquad (2)$$

As should be expected, $R^* \to 1 - N/X$ for large test sets.

We will use the following equality, taken from [MART-82]:

$$R \sim Beta (\alpha, \beta) =>$$

$$\frac{\beta}{\alpha} \frac{R}{1 - R} \sim F (2\alpha, 2\beta) \qquad (3)$$

From this, we get the transformation

$$R = \frac{\alpha F}{\beta + \alpha F} \qquad (4)$$

Thus, the lower and upper bounds for R* are given as

$$R_l^* = \frac{\alpha}{\alpha + \beta F_{1-\gamma/2} (2\beta, 2\alpha)} \qquad (5)$$

$$R_u^* = \frac{\alpha F_{1-\gamma/2} (2\alpha, 2\beta)}{\beta + \sigma F_{1-\gamma/2} (2\alpha, 2\beta)} \qquad (6)$$

For the posterior distribution, we can use the same expressions but replacing α by $\alpha+X-N$ and β by $\beta+N$. When $N = 0$, (5) gives us, for the lower limit:

$$R_l^* = \frac{\alpha + X}{\alpha + X + \beta F_{1-\gamma/2} (2\beta, 2\alpha + 2X)} \qquad (7)$$

Note that $R \sim U(0,1) \equiv R \sim Beta(1,1)$. For this special case (7) is reduced to

$$R_l^* = \frac{1 + X}{1 + X + F_{1-\gamma/2} (2, 2 + 2X)} \qquad (8)$$

THE MODEL OF SOISTMAN AND RAGSDALE

The model developed by Soistman et al. (1985) for describing the effect of quality assurance activities is simple to understand and is intuitively appealing. It uses three sets of factors:

$\{C_i\}$, which describes the software system and its environment

$\{A_i\}$, which describes the error Avoidance techniques used during production

$\{D_i\}$, which describes the error Detection techniques used during production.

The effects of the above mentioned factors are split four-ways in order to cater to the observation that a detection or avoidance technique will have different impacts on different types of statements in a software system.

The current model uses four different statement types, namely interphase, logic, I/O and computation. This gives us the basic formula

$$R = \sum_i C_i \frac{A_i}{1 - D_i(1 - Ai)} \qquad (9)$$

The values of $\{A_i\}$, $\{C_i\}$ and $\{D_i\}$ are found through the use of check lists. Each factor can be computed directly from numbers in the check lists. Generally:

$$A_i = 1 - \prod_j a_{ij} \qquad (10)$$

$$D_i = 1 - \prod_j d_{ij} \qquad (11)$$

$$C_i = \frac{1}{n} \sum C_{ij} \qquad (12)$$

We will use the notation a_i for the first product and d_i for the second one, so that (9) can be changed to

$$R = \sum_i C_i \frac{1 - a_i}{1 - a_i(1 - d_i)} \qquad (13)$$

It is interesting to note how R_i varies with changing a_i, d_i when one of them are kept constant. This is shown in the following two figures:

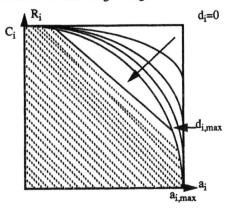

Figur 1: $R_i(a_i)$ for fixed d_i

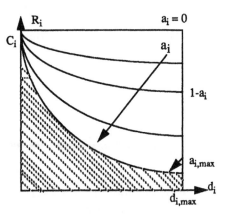

Figur 2: $R_i(d_i)$ for fixed a_i

Note that the upper limit of R occurs when $a_i = d_i = 0$ for all i. This gives

$$R_{max} = \Sigma \, C_i$$

From figures 1 and 2 we can make some interesting observation. Let us first make some simplifications. Define α and δ as follows:

$$\tilde{\alpha}_i = (\prod_j a_{ij})^{1/|A|} \qquad (14)$$

$$\tilde{\delta}_i = (\prod_j d_{ij})^{1/|D|} \qquad (15)$$

This allows us to write

$$\tilde{a}_i = \tilde{\alpha}_i^{|A_i|}, \quad d_i = \tilde{\delta}_i^{|D_i|} \qquad (16)$$

where $|A_i|$ and $|D_i|$ are the number of avoidance and detection techniques used.

We see that

1. Avoidance techniques are much more important than detection techniques.

2. An increase in the number of avoidence techniques will in general have a larger impact on R_i than an increase in detection techniques.

3. An exception to 2) occurs when we already have employd a large amount of detection techniques.

(Soistman, 1985) uses this model to compute R for each component in a software system. These reliabilities are then inserted into Cheungs Markov model and used to study the system reliability. See Cheung (1980).

We will, however, use it to study the impact of different detection and avoidance techniques on the systems reliability and the project costs. This is the topic of the next chapter.

A BETA MODEL FOR THE NUMBER OF TESTS

As mentioned earlier, we can use our R estimate to build a Beta prior distribution that describes our current confidence in the system reliability. This can be done in the following way

1. Compute $R_{nominal}$ an R_{worst} by using the formula shown earlier. The a_i's and b_i's for the two R estimates are found in check list in [SOIS-85].

2. The following relation holds:

   $$R_{nominal} = R_{worst} + \sigma_R$$

3. The computed σ_R only reflects the uncertainty in the effects of the avoidance and detection techniques. Experience showns that the uncertainties due to different personell is of the same order of magnitude. We will thus use $\sigma = 2\sigma_R$.

4. The tables C_1 and C_2 in [MART-82] uses the nominal value (expected vlaue) and the lower or upper 95 percentice. This is equivalent to 1.65 σ, when we use the normal approximation. We will thus use the following tuple for the table C_2 in Martz et al., (1982):

   $$\{R_{nom} - 3.3\ \sigma_R, R_{nom}\}$$

   This will give us values for α and $\alpha+\beta$ in the distribution Beta (α,β).

The whole software production process can be described by the following diagram:

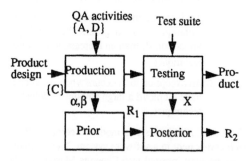

Figure 3: Relationship between QA, testing and the modelling process

We will, on the average, observe $X(1-R_n)$ failures. Our decision rule is then

"If $2C_X X + C_C X(1-R_n) < Cq$, then validate through testing, else introduce the extra quality assurance activity".

C_x will depend on the tools available for

- generating test cases

- bringing the system into the right state for the execution of the test

- checking the test results against the expected results

Our experience is that both C_x and C_c easily can be 1 person-day each. The cost Cq will depend on the type of quality assurance activity.

A MEAN VALUE MODEL

Let us define X, R_i, R_j by the equations

$$R_i = \frac{\alpha}{\alpha+\beta}, \quad R_j = \frac{\alpha+X_{ij}}{\alpha+X_{ij}+\beta} \qquad (17)$$

For $\alpha + \beta \gg 1$, we also have that

$$\alpha \approx \frac{R_i^2}{\sigma_i^2}(1-R_i) \qquad (18)$$

By combining (10) and (11), we get

$$X_{ij} = \left(\frac{R_i}{\sigma_i}\right)^2 \frac{1-R_i}{1-R_j}(R_j - R_i) \qquad (19)$$

It is important to note that X_{ij} does not depend only on the reliability difference $R_j - R_i$ but also on R_i and R_j alone plus the variance of R_i. Thus, X_{ij} will also depend on the number of quality assurance activities and not just on the achieved reliability. Furthermore, it follows that no quality assurance activity has a fixed value measured by the number of tests. The value will depend on how we arrived at the current quality assurance based reliability estimate.

As for the general Bayes theory we see that if we are 100% sure that our prior is the right value, the $\sigma_i = 0$ and $X_{ij} = \infty$ for any R_i, R_j excepted when $R_i = R_j = 1$.

PUTTING THE MODELS TO USE

As our example we will use a medium system built at DELAB. It is characterized by the following features:

Predominantly intractive, single operational misson, many independent operations required, minimal hardware interface, extensive software interface, extensive human interface and a wide range of error-prone inputs.

These characteristica gives us the following C values:

$$\{C_i\} = \{0.305, 0.302, 0226, 0.159\}$$

The basic set of detection techniques used is the following:

Infrequent peer walkthroughs, infrequent quality audits, preliminary design reviews, informal unit testing and formal qualification testing.

This set of techniques gives us the following d values:

$$\{d_i\} = \{0.179, 0.218, 0.241, 0.206\}$$

The basic set of avoidance techniques suggested for this project was the following:

use of a software support library, high order language, hierarchical top down design, and structured code.

This gives us the following values:

$$\{a_i\} = \{0.204, 0.235, 0.204, 0.240\}$$

When we insert these values into the reliability formula, we get $R \approx 0.932$. This was considered to be much too low.

We will consider the following alternatives

1)   Add more avoidance techniques

2)   Add more detection techniques.

In addition, one could consider changing the system or its environment, but these alternatives are not considered here. We will compare the following alternatives:

a)   Move from infrequent to frequent quality audits and from infrequent to frequent peer walk throughs

b)   Buy and use a structured analysis tool.

These two alternatives have the following impacts:

a) $\{d_i\} = \{0.108, 0.138, 0.153, 0.135\}$
      This gives us $R \approx 0.953$

b) $\{a_i\} = \{0.135, 0.163, 0.189, 0.178\}$
      This gives us $R \approx 0.953$.

It is thus reasonable to assume that the two alternatives are equivalent as far as reliability is concerned.

The important question is: Which alternative is better from an economic point of view.

Frequent audits and walk throughs are defined as weekly audits and walk throughs. The extra cost this adds to the project must then be compared to the cost of using the new structured analysis tool. We may have to include aquisition costs, learning costs and the time needed to use the tool in an efficient manner.

The purpose of this paper is not to do a cost-benefit analysis of the two alternatives. The main reason for this is that the costs will depend on a large amount of factors such as the availability of personnel to do walk throughs and quality audits, the cost and quality of the structured analysis tool and so on. It is, however, possible to use the shown method to compare alternatives and perform a cost-benefit analysis under the border conditioins available to each organization.

THE IMPACT ON THE SIZE OF THE TEST SUITE

A different way to look at the process is to consider the alternative of confidence through testing. We can then use our available information in the following way:

For the first set of quality assurance activities we found $R_{non} = 0.932$. In the same way we finds $R_{worst} = 0.886$. This gives us $\sigma_R = 0.046$ and thus the tuple $\{0.78, 0.93\}$. From the C_2 table mentioned earlier we find $\alpha = 7.95$ and $\alpha + \beta = 8.37$.

The number of successful tests needed to obtain a $1-\gamma$ percentile lower limit can be found from (7)

$$X = \frac{R_{low}}{1 - R_{low}} \beta F_{1-\gamma}(2\beta, 2\alpha + 2x) - \alpha \quad (20)$$

If we assume $X > 60$ then the following approximation holds:

$$X \approx \frac{R_{low}}{1 - R_{low}} \beta F_{1-\gamma}(2\beta, \infty) - \alpha \quad (21)$$

Let us compare the size of the following two test suits:

X_1: the number of tests neede to assume $R > 0.99$ with 95% confidence when we assume no prior information

X_2: the number of tests needed to assure $R > 0.99$ with 95% confidence when we include the prior informatioin obtained through the quality assurance

For the no-information case, we asssume that $R \sim [0,1]$. Thus

179

$$X_1 = \frac{0.99}{0.01} F_{0.95}(1, \infty) - 1 \implies X_i = 296$$

$$X_2 = \frac{0.99}{0.01} 0.42 F(0.84, \infty) - 7.95 \implies X_2 = 152$$

We have thus saved 144 test cases, at least. If our reliability goal had been 0.999, we would get $X_1 \approx 2995$ and $X_2 \approx 1251$, thus saving 1744 testcases or circa 58% of the tests.

It would now be tempting to find the Beta distribution for the improved process i.e. inclusioin of the structured analaysis tool. It should then be possible to compute a new number of necessary tests, X_3, and find out how many tests this extra quality assurance activity saved us.

Here we run into a problem. Not only is the R_{nom} higher (0.953) but the σ_R is smaller. Since a smaller σ_R indicates higher confidence, it also follows that we need mor test results to move the estimate away from its expected value.

In my opinion, we should use the same σ_R when we compare alternatives. If we do this, we get $X_3 = 123$ test cases. This indicates that the structured analysis tool saves 29 test cases.

We can, alternatively, use the mean value model. This model turns out to be more robust than the stright Beta model and thus more useful.

Let us use the same examples as we used previously in this chapter:

1. $R_1 = 0.932, \sigma_R = 0.046, R_2 = 0.99$

$$X_{12} = \left(\frac{0.932}{0.046}\right)^2 \frac{0.068}{0.01} 0.067 \Rightarrow X_{12} = 187$$

It is reasonable that we get lower values here since we are considering the mean values instead of the lower bound of R.

2. $R_1 = 0.953, \sigma_R = 0.035, R_2 = 0.99$

$$X_{12} = \left(\frac{0.053}{0.035}\right)^2 \frac{0.047}{0.01} 0.046 \Rightarrow X_{12} = 160.$$

Note that now we get a lower X_{12} estimate, as we should expect

3. $R_1 = 0.932, \sigma_R = 0.046, R_2 = 0.953$

$$X_{12} = \left(\frac{0.032}{0.046}\right)^2 \frac{0.068}{0.047} 0.021 \Rightarrow X_{12} = 12$$

This is thus the "value" of adding the use of a structured analysis tool to the project or -

alternataively - change review and walkthrough frequency to once a week (frequent).

CONCLUSION

This paper shows that the quality assurance model of Soistman and Ragsdale can be used to

- provide information on the current level by confidence, which can be described by a Beta prior distribution of the systems reliability R.

- compare alternative sets of quality assurance activities and thus help the project planner to select the cheaper quality assurance activity set for a predefined level of reliability.

By comparing the reliability increase caused by extra quality assurance activities with the number of fault-free tests needed to achieve the same reliability level, it is possible to find the "value" of each quality assurance acitivity, measured by a test equivalent.

REFERENCES

Parnas, D., Shouwen, A.J. and Kwan, S.P., (1990) Evaluation of Safety-Critical Software, Communications of the ACM, vol. 33, no 6, 636-648

Martz, H.F. and Waller, R.A. (1982), Bayesian Reliability Analysis, John Wiley & Sons, 1982

Soistman, E.C. and Ragsdale, K.B., (1985) Impact of Hardware/Software Faults on Systems Reliability. Procedures for Use of Methodology Rome Air Development Center, Griffiss Air Force Base, NY RADC-TR-85-228, vol II (of two)

Cheung, R.C. (1980), A User-oriented Software Reliability Model, Transactioins on Software Engineering, vol. SE-6, no. 2, 118-125

AUTHOR INDEX

AUTHOR INDEX

KEYWORD INDEX